great inventions
we take for granted

clive gifford

great inventions
we take for granted

clive gifford

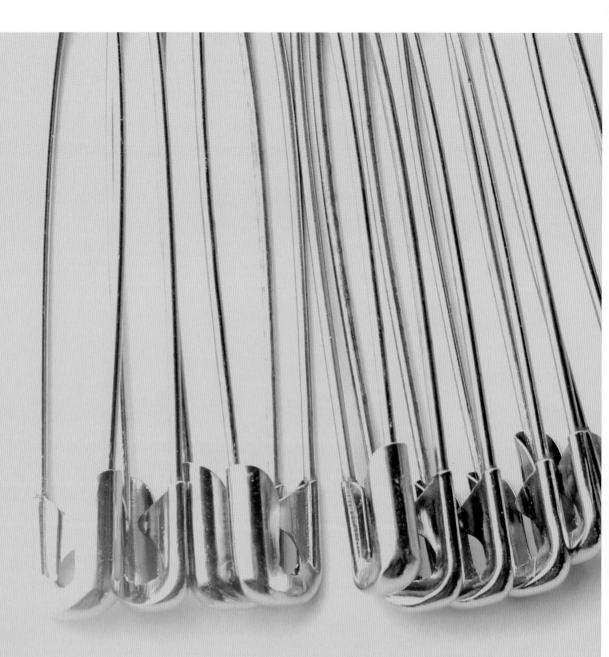

© 2019 Moseley Road Inc.
www.moseleyroad.com

All rights reserved. No part of this publication may be reproduced, stored in a retrieval system, or transmitted, in any form or by any means, electronic, mechanical, photocopying, recording, or otherwise, without prior written permission from the publisher.

President: Sean Moore
Production Director: Adam Moore
Editorial Director: Lisa Purcell

Original book concept and cover by Adam Moore
Art Direction: Duncan Youel at OilOften, London. www.oiloften.co.uk
Book Design: Nicola Plumb
Picture Research: Frances Beard

ISBN: 978-1-62669-155-1

Printed and bound in China

10 9 8 7 6 5 4 3 2 1

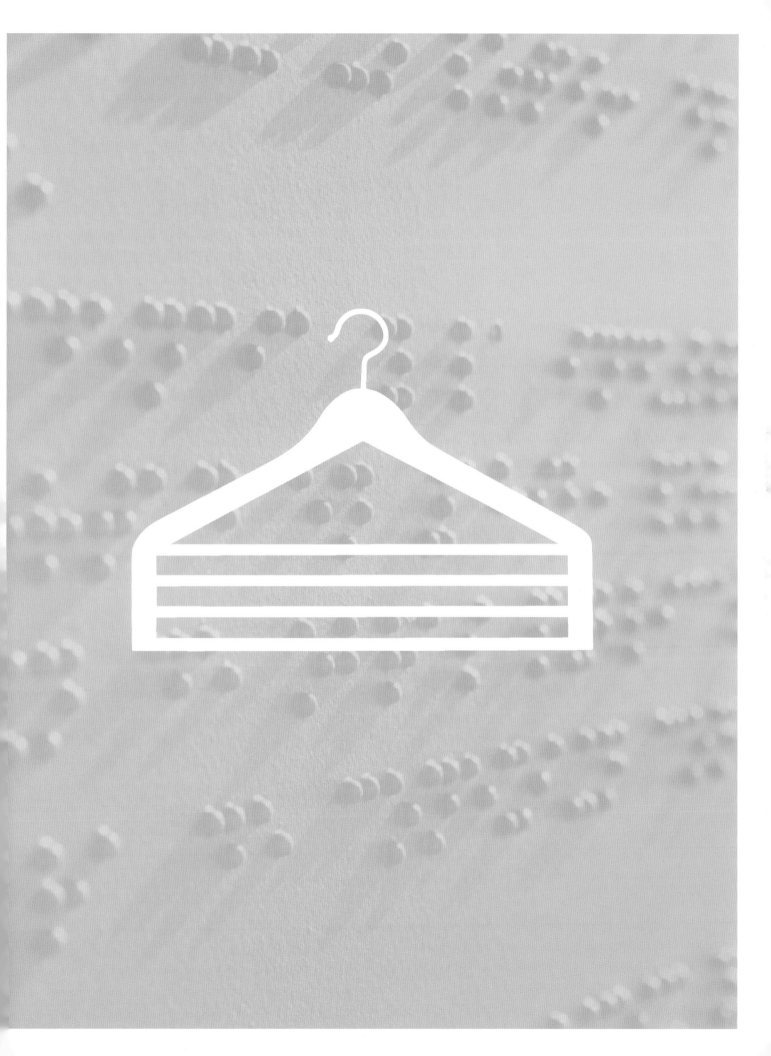

contents

No.1	the band aid	10-11		**No.21**	the can and can opener	50-51
No.2	the candle	12-13		**No.22**	the cotton swab	52-53
No.3	the egg carton	14-15		**No.23**	the dishwasher	54-55
No.4	the tetra pak	16-17		**No.24**	the electric toaster	56-57
No.5	the microwave oven	18-19		**No.25**	the safety pin	58-59
No.6	the mouse trap	20-21		**No.26**	the sewing machine	60-61
No.7	sliced bread	22-23		**No.27**	toilet paper	62-63
No.8	the tea bag and ice tea	24-25		**No.28**	velcro	64-65
No.9	the vacuum cleaner	26-27		**No.29**	spectacles	66-67
No.10	the zipper	28-29		**No.30**	super glue	68-69
No.11	the toothbrush and toothpaste	30-31		**No.31**	central heating	70-71
No.12	breakfast cereal	32-33		**No.32**	the safety razor	72-73
No.13	the iron	34-35		**No.33**	the cellular mobile phone	74-77
No.14	aluminum foil	36-37		**No.34**	soap	78-79
No.15	plastic wrap	38-39		**No.35**	the refridgerator	80-81
No.16	disposable diapers	40-41		**No.36**	frozen food	82-83
No.17	coat hangers	42-43		**No.37**	the fork	84-85
No.18	antibiotics	44-45		**No.38**	the safety match	86-87
No.19	aspirin	46-47				
No.20	braille	48-49				

home

No.39	bubblewrap	90-91
No.40	the banknote	92-93
No.41	the computer mouse	94-95
No.42	the paperclip	96-97
No.43	the sandwich	98-99
No.44	the barcode	100-101
No.45	the wheel	102-103
No.46	the thermos flask	104-105
No.47	the wheelbarrow	106-107
No.48	the elevator	108-109
No.49	cat's eyes	110-111
No.50	the lawn mower	112-113
No.51	the magnetic compass	114-115
No.52	the umbrella	116-117
No.53	traffic lights	118-119
No.54	the seat belt	120-121
No.55	the adjustable wrench	122-123
No.56	the power drill	124-125
No.57	pens and pencils	126-129
No.58	the post-it note	130-131
No.59	the stapler	132-133
No.60	the watch	134-137
No.61	the postage stamp	138-139

work

No.62	the bikini	142-143
No.63	the tv remote	144-145
No.64	the whistle	146-147
No.65	the swiss army knife	148-149
No.66	the hammock	150-151
No.67	headphones	152-153
No.68	lipstick	154-155
No.69	the newspaper	156-157
No.70	the little black dress	158-159
No.71	the credit card	160-161
No.72	computer games	162-165
No.73	the website	166-167
No.74	the digital camera	168-169
No.75	fun and games	170-173
No.76	the radio	174-177
No.77	the bicycle	178-181
No.78	party time	182-185
No.79	sound recording and playback	186-191
	credits	192

play

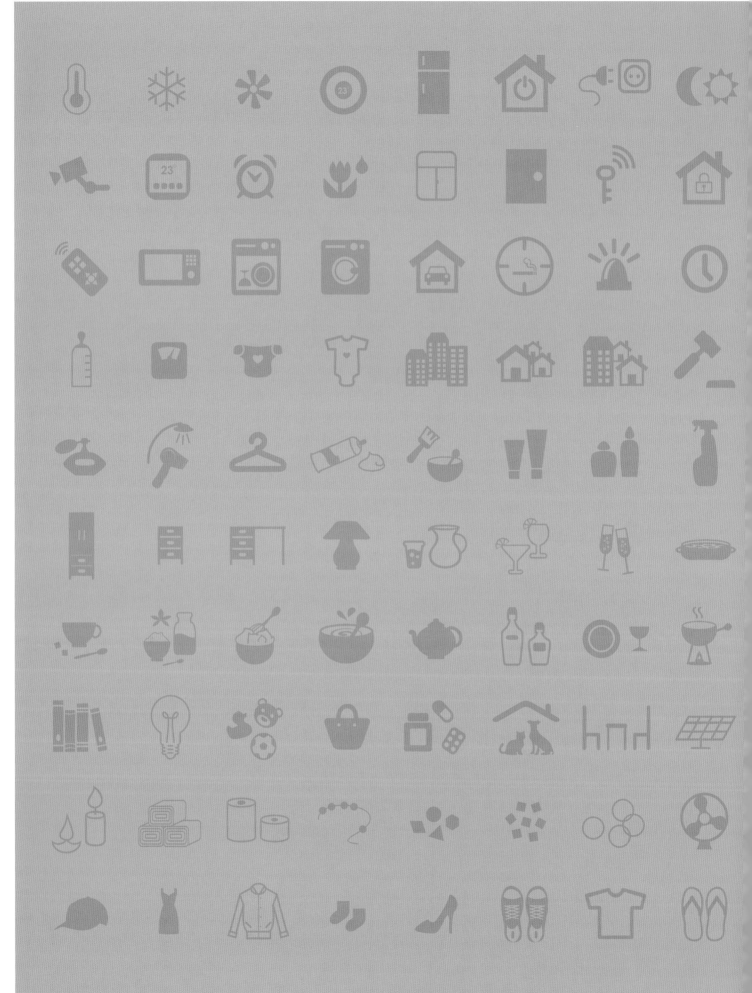

home

Great Inventions We Take For Granted

the band aid

An absolute staple of any home's first aid kit, the personal sticking plaster or adhesive bandage came about through the simple act of one husband seeking to help out his wife.

Great Inventions We Take For Granted

No.1 the band aid

Dickson was rewarded for his invention with a promotion to vice president, until his retirement in 1957, a year after the first decorated bandages for kids – BAND-AID Stars 'n Stripes were launched with bright primary color designs. Twelve years after his retirement, sticking plasters reached the Moon as part of Apollo 11's first aid kit.

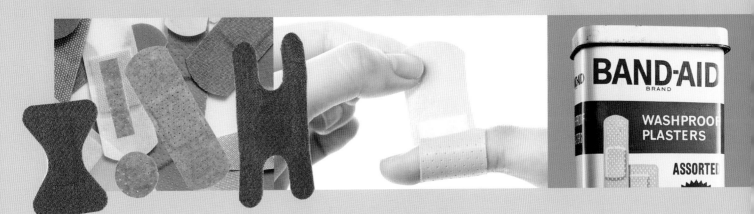

Earle Dickson worked as a cotton buyer for large medical and healthcare firm, Johnson & Johnson in New Brunswick, New Jersey. In 1917, he married Josephine Francis Knight and by 1920 had got used to the regular routine of bandaging the occasional cut, burn or graze on his wife's hands and arms that were common in early 20th Century housework. Available bandages at the time were often big and bulky and hard to apply oneself with tape over a lint or gauze dressing.

Dickson decided to cut up a number of pieces of sterile gauze and placed them at intervals on the centre of strips of surgical tape. He folded the gauze over into a narrow pad and added a band of crinoline so that the surgical tape wouldn't stick to itself. The result meant that his wife had easy access to a roll of pre-prepared dressings which she could cut a small piece off at any time to apply a small bandage to a knife-nicked finger or a grazed knuckle. For the first time, a dressing could now be applied easily and one-handedly by the injured person without assistance, a fact not lost on Johnson & Johnson executives when Dickson finally showed them his innovation.

The new plasters went into production in 1920 at Johnson & Johnson's mills but lacking a product name. Executives struggled to come up with a title for the new product until the superintendent of the mill, W. Johnson Kenyon, had a spark of inspiration and suggested BAND-AID® - now a globally recognised brand.

Sales, however, did not skyrocket from day one. In the first year of production, only $3000 of the product was sold – a poor return on the investment and possibly down to the product sizing as the bandages came in large, relatively unwieldy sections measuring two and a half inches long and 18 inches wide. Changes were made on both the marketing and product design sides. By 1924, the plasters were being produced in a range of convenient sizes – particularly the three inches long, three-quarters of an inch wide plaster people are most familiar with today, and the hallmark red thread pulled to easily open each plaster's packaging was introduced. Johnson & Johnson hired travelling salesmen to demonstrate how to use the product and also distributed large numbers of BAND-AIDs for free amongst Boy Scout troops across the entire United States. Take-up rose sharply and further innovations such as the first completely sterile bandages in 1939 and clear strip bandages in 1957 have seen more than 100 billion of this brand of plasters produced as well as millions more from many imitators and rival companies.

Above left Modern plasters come in a variety of shapes, sizes and materials, from clear plastic to fabric dressings.
Above center The plasters' simple design allows easy application with just one hand.
Above right The famous BAND-AID® tin became a design icon and, once empty, a store for all sorts of household nick-nacks.

The inventor of the first candle is lost to history, but for over 5,000 years, this humble device has shined a light on an otherwise largely dim world. Despite the ubiquity of electric light, candles are still widely used in ceremonies and rituals. People also turn to candles for power cuts, emergencies and for setting a mood. Boosted by the fashion for burning scented candles, the US candle industry is still worth over $2 billion a year.

the candle

The very first candles may have been a simple lump of animal fat dropped in a fire. Ancient Egyptians were dipping reed rushes in melted animal fat to make candle torches by 3000BCE. Similar simple candles were used by other early cultures whilst the Romans started dipping twine into wax made from tallow derived from cows and sheep by 400-500BCE. The twine formed a flammable wick inside the hardened tallow.

Tallow had a strong odour that many found unpleasant due to its high glycerine content. Beeswax was discovered, possibly first in China, to burn better than tallow but giving off a sweet, pleasant smell in addition. Beeswax candles became popular in churches and the homes of the wealthy, as it remained expensive throughout the Middle Ages. Chandlers (candle makers) would produce candles by repeated dipping of the wick into the tallow or beeswax until moulds were invented to pour the molten wax into. In 1834, Joseph Morgan from Manchester, England invented a machine that mechanized the production of moulded candles using a moving piston to eject the candles after they solidified at a rate of up to 1,500 per hour.

Whale Of A Time
Tallow candles remained common until the growth of the whaling industry in the late 18th century. Spermaceti, a wax obtained by crystallizing sperm whale oil began to be produced in large quantities. Spermaceti gave off little odour, just like beeswax, but was cheaper and usually burned with a brighter light than other waxes. Further competing waxes came from industrial chemistry in the form of processing stearic acid and, particularly, paraffin wax which was isolated from petroleum for the first time in 1830 by Carl Ludwig von Reichenbach and became the most popular form of cheap household candle.

Candles On Cakes
Historians believe that the Ancient Greeks made offerings of cakes with candles to represent the Moon to the temple of Artemis, the Greek goddess of the moon and hunting. Candles on birthday cakes became a popular celebration for children called Kinderfesten in 18th Century Germany with each candle representing a year of the child's age, plus an additional candle in the hope they would live another year.

No.2 the candle

"There was cake as large as any oven could be found to bake it, and holes made in the cake according to the years of the person's age, everyone having a candle stuck into it, and one in the middle."

Andrew Frey describing a birthday party for German nobleman, Count Ludwig von Zinzendorf in 1746.

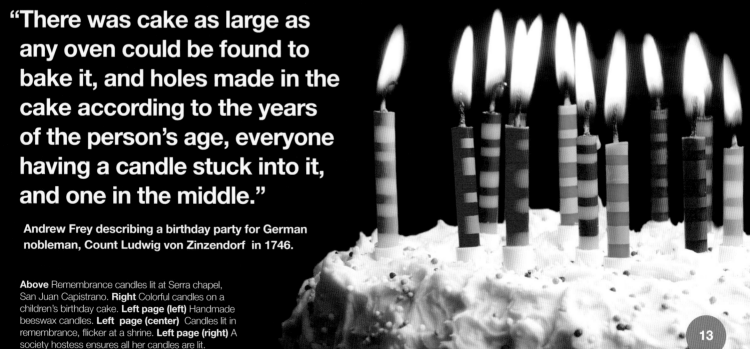

Above Remembrance candles lit at Serra chapel, San Juan Capistrano. **Right** Colorful candles on a children's birthday cake. **Left page (left)** Handmade beeswax candles. **Left page (center)** Candles lit in remembrance, flicker at a shrine. **Left page (right)** A society hostess ensures all her candles are lit.

Great Inventions We Take For Granted

the egg carton

Americans are big egg eaters, consuming an average of 274 eggs per capita in 2017 alone, according to the US Department of Agriculture.

Whether fried or scrambled for breakfast or used in quiches and other baked recipes, eggs are wonderfully versatile food sources, but they can be vulnerable unless packaged well. For centuries, eggs were transported in baskets all together, but sometimes cushioned with layers of straw. Breakages were inevitable and common. In 1906, a Liverpool, England man, Thomas Peter Bethell produced the first of a number of versions of his Raylite Egg Box. This sturdy wooden chest had card or wood frames inside that sectioned off and protected each egg. The Raylites were designed to be re-used and were not suitable for stacking on store shelves.

Five years later across the Atlantic, Joseph Coyle overheard a conversation in a hotel near to the offices of the newspaper, Interior News, he published in Bulkley Valley in British Columbia. The pair were a hotelier and a farmer arguing over the delivery of eggs with many broken shells due to their transportation by horse and cart over rough ground. Intrigued and curious, Coyle set about designing his own solution, coming up with a design he patented in 1918, and the following year, selling the newspaper, to set up businesses selling his Coyle Safety Egg Carton advertised as, "It saves you more than its cost". This easily-assembled box made, initially from newspaper, and later, cardboard, had separate inserts for each egg that kept them safe and in their place.

Egg cartons differed little from Coyle's design until the arrival of a new design using recycled cardboard and paper pulp by an Englishman, H.G. Bennett in 1952. Plastics came late to the packaging party but were notably employed by Dow Chemicals who along with young entrepreneur (and now billionaire businessman) Jon Huntsman, patented the first egg carton made out of plastic foam in 1967.

Above A ten egg carton found in supermarkets. **Right page (left)** Stackable designs protect layers of eggs. **Right page (center)** Styrofoam egg carton. **Right page (right)** Empty paper-based cartons destined for recycling.
Right page (below right) A page from Joseph Coyle's original patent, filed in 1927.

No.3 the egg carton

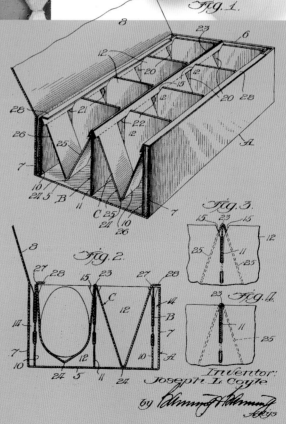

"Is there a more humbly immaculate commercial packaging object than the plain pulp egg carton? This biscuit-colored, biodegradable vessel is a masterpiece of streamlined efficiency."

Calum Marsh, critic, writing in the National Post, 2018.

Great Inventions We Take For Granted

"Pioneered by Tetra Pak, aseptic technology is the most important food science innovation of the 20th century"

Institute of Food Technologists, Chicago

the tetra pak

In 1929, Ruben Rausing and Erik Åkerlund formed the first specialised factory for producing food packaging, in Scandinavia. Based in Sweden, Åkerlund & Rausing became one of the largest packaging manufacturers in Europe

In the 1940s, Erik Wallenberg - a young engineer working for the company - developed a new packaging process to produce milk containers from a continuous roll or sheet of paper, first rolled into a cylinder and then folded from two different sides to creature four triangular faces – a tetrahedron. The resulting containers were cheaper to produce and transport than heavier glass bottles. Wallenberg was paid six months wages as a bonus but it was Rausing's name on the patent when it was filed. Later versions contained a sandwich of paper, plastic and aluminium to create a sterile, aseptic system for filling containers with milk preventing it from coming into contact with harmful microorganisms present in air.

In 1952, the first Tetra Pak machine that could make the cartons was delivered to a dairy in Lund called Lundaortens Mejeriförening. It was used to pack cream into small 3.52 fl oz (100ml) packages. Larger cartons for milk were manufactured from 1954 onwards with the first Tetra Pak machine exported out of Sweden to a dairy processor in Hamburg, Germany, the same year.

Tetra Paks developed into different designs and shapes including the Tetra Brik in 1963, capable of holding juice, soup and other foodstuffs. The company, now with its headquarters in Switzerland, is today the largest supplier of packaging in the world with revenues of over US$12 billion. The company produces somewhere in the range of 180 billion packs a year – more than 20 for every single member of the human population. Some 500 million packs are consumed every day in nearly every country in the world.

No.4 the tetra pak

Left page Erik Wallenberg with the tetrahedron package.
Above top left Tetra Pak advert with Tetra Classic, Italy, 1960s.
Above top right Tetra Brik advert, 1960s.
Left Various shapes and sizes of Tetra Pak products.
Above Production line of Tetra Paks in Malaysia.

Great Inventions We Take For Granted

"He had a reputation for thinking outside the box. He was totally driven to create new things and had an intuitive sense for how things worked."

Rod Spencer Jr, grandson of Percy LeBaron Spencer

the microwave oven

Whilst our ancestors may have struggled over a wood-lit fire for hours, 'heat in a hurry' is now an everyday circumstance with microwave ovens, following the accidental discovery of the heating power of microwaves.

These short wave examples of energy on the electromagnetic spectrum, between infrared and radio waves, can pass through many foods, agitating molecules to produce heat.

Percy LeBaron Spencer was a renowned electrical engineer working at Raytheon on components for use in early RADAR systems, specifically cavity magnetrons - devices first invented by John Randall and Harry Boot in 1940 which could generate microwaves. One day in 1945 whilst standing close to an operating magnetron, Spencer found the peanut cluster bar (or chocolate bar as the story varies) in his pocket warming up. Intrigued, he and his team performed a series of food experiments over the forthcoming days, finding that raw popcorn 'popped' excitedly and eggs exploded hazardously when experiencing the full force of the magnetron's energy.

Spencer filed a patent for a cooking device based on his discovery and Raytheon began developing a practical oven, first tested out in a restaurant in Boston. The original Radarange microwave oven was taller than the average man, as wide as a large refrigerator and at around 750 pounds, heavier than both. The magnetron required a water coolant system so the entire oven needed to be plumbed into a mains water supply.

Home On The Range / Home Ovens

By 1955, Raytheon had begun licensing its microwave technology, and the first microwave oven designed for consumers went on sale from the Tappan Stove company of Ohio. The Tappan RL-1 was mounted in a kitchen wall and cost US $1,295 (over $11,000 today), putting it out of the reach of the pockets of ordinary people. Raytheon subsidiary Amana began selling household Radarange countertop ovens in 1967 but even by 1970, as designs shrank in size and price, only 40,000 ovens were sold in the US. It was the 1970s and 1980s where microwaves took hold, selling in their millions, in part spurred on by microwaveable food innovations such as popcorn (Spencer had taken out a patent in the 1940s on microwaveable popcorn in a bag but featuring an entire bagged corn kernel), french fries and especially, microwave-suitable TV dinners and ready meals. Now, more than 90 per cent of all American households own a microwave.

Today, microwave ovens vary enormously in size from giant ovens used in industry to cure composite materials to small models that can fit in a car or on a worker's desk. The iWave Cube, launched in 2008, measured just 10.5 inches square and 12 inches tall whilst the Wayv Adventurer, launched in 2017 for microwaving on the move, is even smaller – a 12 by 5 inch flask weighing just two and a half pounds.

No.5 the microwave oven

Percy Spencer received no royalties for his discovery and invention but Raytheon did pay him two dollars – the standard fee from the company to any of its employees that produced a patentable invention.

Left hand page Percy LeBaron Spencer.
Above left Microwaveable popcorn – ready in a minute.
Above center A modern slimline consumer microwave oven.
Above right Micro Cupol microwave oven, designed in 1969 by Carl-Arne Breger, Husqvarna, c.1973.
Below Dial control of cooking duration.

How They Work
A home microwave oven uses mains electricity, stepped up by a transformer to 3,000 to 4,000 volts, to power a magnetron. This device uses interacting electric and magnetic fields to produce microwaves which oscillate and reverse their electric field 2.45 billion times each second. The microwaves are directed by a wave guide into the oven's cooking chamber – a sealed metal box – where they bounce around, striking and exciting molecules in food. Water molecules in food feature a negatively-charged oxygen atom and hydrogen atoms which are positively charged. The molecules turn to align with the polarity of the microwaves' electric field. This field changes its polarity billions of times each second, causing the molecules to constantly flip back and forth, generating heat through friction as they rub together. The end result is greatly reduced cooking times – sometimes 6-10 times shorter than in a conventional electric or gas oven.

Great Inventions We Take For Granted

the mouse trap

Solutions, besides having a good mouser as a cat, have long been sought to those individual single ounces of rodent trouble that cause alarm, nibble foods and gnaw through wiring in millions of homes.

Early American homesteaders made their own mousetraps, as simple as snares, caged baskets or hinged wooden doors dropping mice down into containers of water or deadfall traps where the mouse's arrival triggers the fall of a heavy weight or a series of lethal spikes. Various engineers including the inventor of the first portable automatic machine gun, Hiram Maxim, turned their hands to inventing their own design of trap. A visit to The Trap History Museum in Galloway, Ohio conveys the great variety of designs - over 1500 different examples of mousetraps can be found there. Continuing innovation in trap design, has seen large numbers of plastic humane traps where the mouse is held for later release into the wild. Yet, the device many turn to when they seek to remove rodent pests plaguing their house or barn is the simple and humble spring-loaded snap trap.

In 1894, Illinois inventor William Chauncey Hooker developed what he described as, "a simple, inexpensive and efficient trap adapted not to excite the suspicion of an animal, and capable of being arranged close to a hole." It was the first commercially successful snap trap and featured a wooden block onto which was fitted a coiled spring which was released when a piece of bait was moved from the block by a mouse or other rodent. The spring snapped a wire called a striker or killing bar over with ferocious speed, snapping the neck of the mouse or trapping its body in place.

A similar design was perfected by John Mast from Lititz in Pennsylvania in 1899. He later formed a company to sell his Victor traps and merged with the Animal Trap Company in 1905 that owned some rights to Hooker's traps. Across the Atlantic, a British ironmonger from Leeds, James Henry Atkinson filed applications for several different inventions, amongst them the Little Nipper mousetrap in 1897. This also used as a spring loaded killing bar which snapped shut fast in just 38 thousandths of a second. Atkinson sold his patent to the Proctor Brothers. This company still manufacture similar Littler Nipper traps to this day, claiming a commanding 60 per cent of the UK market whilst Victor traps are still sold in their millions each year in the United States.

No.6 the mouse trap

With over 4,000 patents assigned, the mousetrap is the single most patented device in the US. The Patent Office still receives an average of 20 new designs for mousetrap inventions every year.

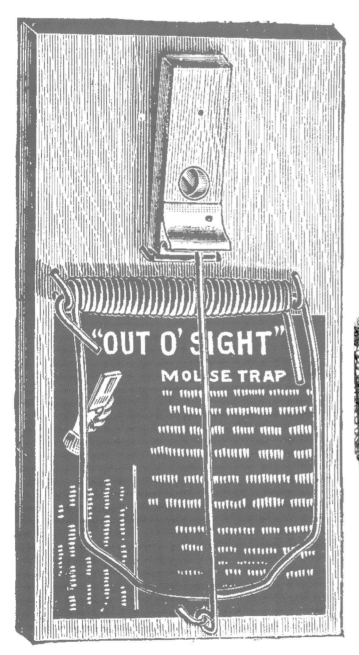

Left page (above) A classic spring loaded trap is cocked and awaiting bait. **This page (left)** Mouse trap advertising from the 19th Century. **This page (above and below)** Examples of different live capture and cage trap designs.

Great Inventions We Take For Granted

> "To protect the cooperating bakeries against the unfair competition of those who continue to slice their own bread...we are prepared to take stern measures if necessary."
>
> John F. Conaboy, the New York Area Supervisor of the Food Distribution Administration, January 26th, 1943.

sliced bread

Bread was bought whole and sliced unevenly by hand until the pioneering invention of a former ophthalmologist and owner of three jewellery stores in the Missourian town of St Joseph, Otto Frederick Rohwedder.

In the 1900s, Rohwedder started tinkering with a design for a machine that automatically sliced and wrapped baked loaves of bread. To gauge the ideal thickness, he placed advertisements in local newspapers offering a questionnaire "for the purpose of determining a thickness of slice which would be most nearly universal in acceptance." It garnered over 30,000 responses from housewives. By 1912, he had a working model in place but struggled to keep the loaf firmly together after slicing, even advocating the use of metal pins running all the way through the bread. A 1917 fire in a factory in Monmouth, Illinois set Rohwedder back a decade as the blaze damaged a factory that had agreed to manufacture his machine. Inside the building were all of Rohwedder's bread-slicing blueprints which were destroyed by the fire. By the time Rohwedder rebounded and perfected his invention, electric pop-up toasters were becoming more popular and the demand for thin, evenly cut bread for toasting was increasing. Rohwedder's machine consisting of multiple blades that separated the loaf into uniform slices before it was wrapped in wax paper to maintain its freshness. Many bakeries were reluctant to invest in the device, fearing that the bread would go stale but the Chillicothe Baking Company of Chillicothe, Missouri, took the chance, putting their first sliced loaf on sale in July, 1928 under the brand, Kleen Maid Sliced Bread.

Within weeks, the bakery's bread sales rocketed by 2000 per cent and others became keen to get in on the act. Wonder Bread was producing sliced loaves by 1930 which greatly popularised the product. No one knows who first came up with the epithet, "the best thing since sliced bread," but American consumers seemed to agree with the phrase. Within five years of Rohwedder's invention launch, most bakeries in the US had bread slicing machines and as much as four-fifths of all bread produced by companies in America was sliced.

No.7 sliced bread

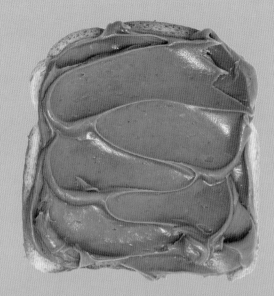

Bread was bought whole and sliced unevenly by hand until the pioneering invention of a former ophthalmologist and owner of three jewellery stores in the Missourian town of St Joseph, Otto Frederick Rohwedder.

The Best Thing Since…Banned!
During World War II, attempts to rationalise and conserve resources saw nations like the UK place public advisory posters advocating a reduction in bread consumption. In the United States, Claude R. Wickard, the head of the War Foods Administration as well as the Secretary of Agriculture, decided to implement a ban on sliced bread, announced on January 18, 1943. The stated aim was to reduce the materials consumed in making the thick waxed paper that the sliced loaves were wrapped in.

Complaints, disapproval and distraught letters from housewives hit the newspapers with one to The New York Times exclaiming, "I should like to let you know how important sliced bread is to the morale and saneness of a household." The ban proved unpopular and the alleged savings not significant enough for it to continue. When the ban was lifted two months later, The New York Times heralded the news with the headline, "Sliced Bread Put Back on Sale; Housewives' Thumbs Safe Again."

Right Word War one poster 'Eat Less Bread' - c1914-1918.
Left Sliced wholemeal bread.
Above Three common toppings – bananas, peanut butter and jam.

Great Inventions We Take For Granted

the tea bag and ice tea

Despite being first grown and drunk in China, Sri Lanka and other parts of Asia, and almost worshipped with due ceremony in Britain and Russia, two of the most recognised tea-related inventions are actually American.

No.8 the tea bag and ice tea

"On any given day, over 159 million Americans are drinking tea. On a regional basis, the South and Northeast have the greatest concentration of tea drinkers."

The Tea Association of the USA

Left page (below) Green tea buds and leaves in a tea plantation field.
Left page (above A pair of tea bags with a tag and string for dunking during infusion. **Above (top left)** A packed box of tea bags.
Above (center left) Colorful 'Arizona' brand ice teas on display.
Left Loose fruit tea. **Above (center)** Pouring hot tea from a teapot.

The first tea plants to reach America are believed to have been brought by French botanist, André Michaux who imported a variety of exotic flora and planted tea plants in Charleston, South Carolina. The state remained the only one in the United States to grow tea commercially in significant quantities. The vast majority of tea was imported from overseas, giving rise to tea merchants in major port cities such as New York and Boston.

One such tea merchant in New York, Thomas Sullivan came up with a neat promotional idea at the start of the 20th Century. He started to send small silk bags containing samples of his loose leafteas to his customers. The only issue was that some of his clients assumed that the tea was to be drunk as is, dropping the bag into a tea pot before applying hot water. Sullivan was quick to react and worked on a design using cotton gauze rather than silk to enable the flavour to escape more easily. Competing tea bag designs, some made of paper rather than cotton and others containing a string for dangling the bag into the pot or cup emerged whilst Boston's William Hermanson patented a paper fibre tea bag that could be sealed using heat, enabling quicker mass production. Early tea bags were small sacks, but since that time, square, round and tetrahedral bags have all been produced. Whatever the shape, Americans in 2017, consumed over 84 billion servings of tea equal to more than 3.8 billion gallons, and much of it produced from tea bags.

Iced Tea

Today, around 80 per cent of all tea consumed in the US is served not hot but ice cold. Many books state erroneously that iced tea was invented at the 1904 World's Fair in St Louis. In truth, it was being drunk and enjoyed decades before. An 1879 cookbook, Housekeeping in Old Virginia by Marion Cabell Tyree, gives a classic sweet recipe with each glass sweetened by two teaspoons of sugar. A recipe, 40 years earlier in The Kentucky Housewife, by Lettice Bryanon was more elaborate involving one and a quarter pounds of loaf sugar, half a pint of cream and a bottle of claret red wine stirred in!

A staggering 19.7 million people attended the Louisiana Purchase Exposition, better known as the World's Fair, in Missouri between April and December of 1904. Visitors wondered at all sorts of exciting new products on display including wireless telephones and brand new foods and beverages including Dr Pepper and puffed wheat cereal. In addition, many pre-existing food industry inventions including waffle cones and peanut butter received welcome publicity boosts at the exposition, including iced tea.

The director of the East Indian Pavilion at the World's Fair, Richard Blechynden, offered free hot tea to visitors, but the scorching summer temperatures meant that take-up was low. He is said to have had a brainwave and began cooling large bottles of his brewed black tea using iced lead pipes. The cooled tea proved far more palatable to visitors and helped popularise a drink to a far wider audience than before. According to the 2017 Zenith RTD Tea Innovation Report, some 11.89 billion gallons of iced tea is expected to be consumed around the world in 2021.

Great Inventions We Take For Granted

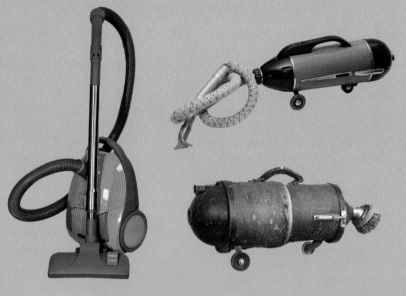

Above Advertisement for General Electric c1950s. **Left** A selection of different vacuum cleaners, old and new. **Far right** Dyson DC07 cyclonic vacuum cleaner. **Right:** A 1950 advert for Hoover's new Model 29 – the first vacuum cleaner the company manufactured in red.

the vacuum cleaner

Prior to the invention of vacuum cleaners, homes tended to be dust-ridden. You might instruct your servants to sweep your rugs and carpets daily, but all this did was move the dust around, not remove it.

Sometimes, carpets and rugs were scrubbed in place, leaving them damp for days at a time or taken outside and beaten and shakenclean. The first notable step forward came with the invention of a 'carpet sweeper' by Daniel Hess from Iowa in 1860. It used spinning brushes to agitate and remove dust from carpets and hand bellows to propel the dust into a container. Chicago native, Ives W. McGaffey patented his "Whirlwind" cleaner nine years later and sold it via a new company, the American Carpet Cleaning Co. It required great dexterity for an operator to turn a hand crank continuously to generate suction whilst pushing the cleaner over the floor.

By Royal Appointment
New cleaning machines in the 1890s pumped air at carpets and cushions, hoping to blow the dust off. British civil engineer, Hubert Cecil Booth, viewed a demonstration of one of these machines at a London music hall in 1901 and pondered why the inventor hadn't reversed the motor to suck rather than blow. Booth's 'Puffing Billy', featured a five horsepower internal combustion engine which generated suction, lifting surface dirt from floors and rugs and carrying it through long pipes into a container inside the machine. The Puffing Billy was fitted to a horse-drawn cart as it was far too big to enter a home. Customers would hirethe machine which would be parked outside and the suction pipes carried into the house via doors and windows. Booth cleverly fitted some of his machines with transparent tubes so customers could actually watch the large amounts of dust and dirt being removed.

Booth's machines became all the rage in high society London, especially after one was used to clean the carpets of Westminster Abbey prior to the coronation of King Edward VII in 1902. The king purchased two of Booth's cleaners outright, one for Windsor Castle, the other for Buckingham Palace. By 1903, London socialites would hire one of Booth's machines for their home, invite their friends, and hold vacuum cleaning parties.

No.9 the vacuum cleaner

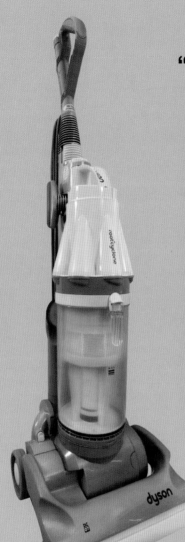

"I'd seen an industrial sawmill, which uses something called a cyclonic separator to remove dust from the air. I thought the same principle of separation might work on a vacuum cleaner. I rigged up a quick prototype, and it did. It took five years of doing nothing but making and testing prototypes. My wife supported us by teaching art. She was wonderful. But most other people thought I was mad."

James Dyson on inventing his first cyclonic bagless vacuum cleaner.

iRobot vacuum cleaner

Janitor's Joy
James Spangler was an American janitor in a department store in Canton, Ohio, who suffered from asthma, not helped by breathing in dust as he swept. In 1907, determined to improve his lot, he built a portable vacuum cleaner using a pillowcase as a dust bag, an electric fan, a tin box and a motor from an old sewing machine. Spangler began manufacturing his 'Electric Suction Sweeper' assisted by his son and daughter at a rate of two to three machines per week. He donated one of his first machines to his cousin, Susan who was impressed as was her husband, William Hoover, who bought all the rights to the machine. The design was improved upon and as sales of Hoover's vacuum cleaners rocketed, the company name becoming synonymous with all vacuum cleaners.

Dyson's Delight
Some regular vacuum cleaners with dustbags can experience a significant reduction in their suction power as the bag gets full of dust and new bags need to be purchased regularly. These inconveniences led British engineer, James Dyson to experiment with bag-free systems. Experimenting tirelessly, Dyson produced a staggering 5,127 different prototypes and versions of his bagless cyclonic vacuum cleaner before the first model, the G-Force, finally went on sale in the mid-1980s. Dyson's vacuum cleaner featured a powerful motor which sucks in dirty air and then spins it around fast. The dirt and dust is flung out of the spinning air and falls to the bottom of the cleaner where it is easily emptied.

Robotic Cleaners
The first robot vacuum cleaner, the Trilobite, was invented by Swedish company Electrolux in 1996. A rival model, the Roomba, launched in 2002 by US company, iRobot, is now the most common robot of any type in the world with over 14 million machines sold. Each robot cleaner navigates itself around a room, using sensors to determine proximity to obstacles such as furniture, avoiding stairs and keeping track of where it has already cleaned before autonomously returning to its power point to recharge its batteries.

Great Inventions We Take For Granted

We have Swedish-American engineer Gideon Sundback to thank for the modern zip fastener. Made of endless different materials and sizes, the zipper has revolutionized the garment industry and many other spheres of everyday life.

the zipper

'While other designers thought of zippers simply as a fastener and tried to hide them, Schiaparelli proudly flaunted them to create visual interest."

TrimLab Fastener and Trim Development Centre, 2014

Chicago-born engineer Whitcomb L. Judson garnered some 30 patents during a 16 year long inventing career following his time serving in the Union Army during the American Civil War. A number of these related to his proposed train service powered by compressed air – the Judson Pneumatic Street Railway, but his 1893 patent for a, "Clasp Locker", developed some three years later, was effectively the first zipper.

Judson's motivation for his invention came from the high boots in fashion at the time which could be tedious and time consuming to don and take off. His device, made of rows of eyed and toothed clasps brought together or pushed apart by moving a guide was designed to "free from the annoyances hitherto incidental to lace-shoes". Judson exhibited his invention for the first time at the 1893 Chicago World's Fair and set up the Universal Fastener Company with two businessmen.

His invention attracted relatively minimal interest and the complexity of design and large size of each clasp element resulted in the fastening unintentionally bursting open on occasion. They also broke too easily. Judson's business endured something of a nomadic existence, moving from Chicago to first Ohio, then Pennsylvania before settling in Hoboken, New Jersey. There, in 1906, the company employed a recently-arrived Swedish émigré, Gideon Sundback, who had trained in Germany as an electrical engineer.

Sundback rose to head designer and in 1909 took out a patent in Germany for his first Separable Fastener before developing the Hookless No.2 design in 1914 which received its US patent three years later. It featured densely packed rows of teeth on two fabric strips offset by half a tooth's height. Each tooth has a projection known as a nib on one side and a dimple depression on the other. When drawn together by the wedges inside the guide, a tooth's nib engages with the dimple of a tooth on the other strip, interlocking them all together – the template of a modern zipper.

First manufactured for boots, money belts and flying suits for the booming aviation industry, a B.F. Goodrich employee is said to have coined the phrase, "Zip 'er up" around 1923. Goodrich registered Zipper as a trademark later that decade. Newly named, the Zipper entered the 1930s where it saw an upturn in its fortunes. Zippers were increasingly used to make easy wear children's clothing whilst the Prince of Wales, heir to the British throne (and in 1936 crowned King Edward VIII) began wearing

No.10 the zipper

According to a report by Global Industry Analysts, the global market for zippers will be $13.7 billion in 2020.

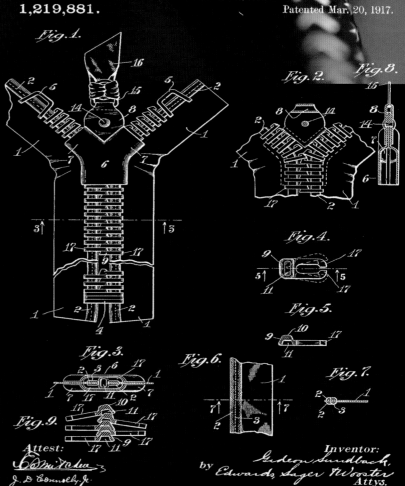

Left Gideon Sundback.
Above A Zipper guide meshes two rows of teeth together.
Right Sundback's 1914 patent application, granted three years later.

trousers with a zipper fly in 1934. This caused a stir in British polite society and leading to a trend amongst the fashionable and dapper that led Esquire magazine in 1937 to praise the zipper for helping men avoid, "the possibility of unintentional and embarrassing disarray."

Perhaps, most influential of all were high end Paris fashion house designers, especially Elsa Schiaparelli who incorporated brightly coloured zipper in her sportswear collections in 1930 and five years later designed evening dresses featuring oversized decorative zippers that were much admired and copied. The Zipper had taken hold and was heavily used in World War II for everything from equipment bags and pouches to pilots and aircrew's flying suits.

First patented in its modern form in 1917

Great Inventions We Take For Granted

"I found that it was not unpleasant. It was painful on my gums and made them bleed as well, but that's not a bad thing and afterwards my mouth felt fresh and clean."

Dr Heinz Neuman on trying the world's oldest toothpaste recipe in 2003.

the toothbrush and toothpaste

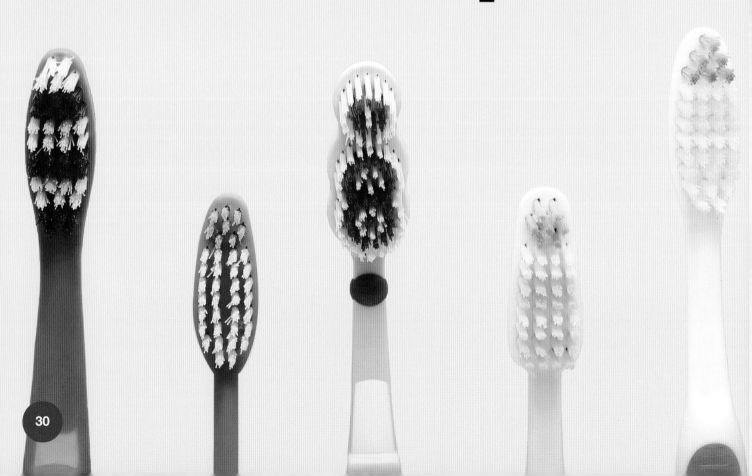

No.11 the toothbrush and toothpaste

According to the American Dental Association, you should spend two minutes brushing your teeth each visit, equivalent to just over 24 hours brushing every year.

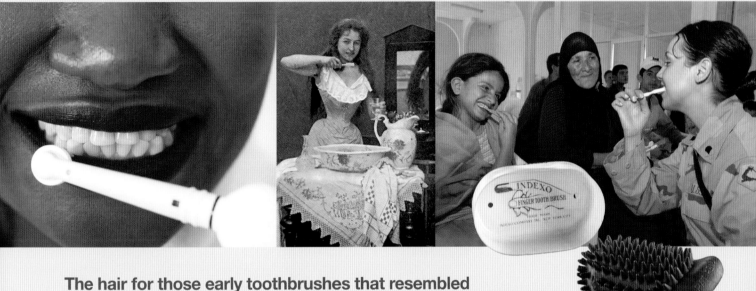

The hair for those early toothbrushes that resembled today's design came from pigs.

Long before there were toothbrushes, there were chew sticks. Ancient Babylonians circa 3,500BCE used to clamp their teeth down on the frayed end of twigs and small branches, chewing the wood to remove food debris and attempt to freshen their breath. This course of action became common in China and the Islamic world where the sticks were called miswak and mostly hailed from the Arak tree (Salvadora persica) whose wood contains substances with some antimicrobial attributes.

The first toothbrushes emerged during China's Tang Dynasty which lasted from 619CE to 907CE. They featured a handle of bamboo or bone embedded with rough hairs from the hides of northern Chinese and Siberian pigs whose bristles were believed to be firmer due to the cold climate they endured. Handmade toothbrushes with pig or boar bristles or softer horsehair began to be imported into Europe in the 17th Century onwards.

Whilst serving a prison sentence in 1780 for instigating a riot in the Spitalfields area of London, Englishman William Addis designed the prototype of the modern, mass-produced toothbrush with a handle carved from a bone saved from a prison meal and some pig bristles obtained from a guard, grouped in rows of small tufts. On release, Addis manufactured and sold this design at great profit; his company, renamed Wisdom, would remain family-owned until 1996. For Addis and the many other manufacturers who sprang up, animal hairs proved the only source of bristles until the invention of nylon in the 1930s. The Weco Products Company of Illinois are believed to have produced the first all-plastic toothbrush with nylon bristles in 1938 which went on sale as Dr West's Miracle Toothbrush.

Toothpaste
Astonishing concoctions including honey and mouse dung, soot, ash from charred trees and, in the case of the ancient Romans, urine were all ingredients in early substances designed to clean the mouth and teeth. The oldest known toothpaste recipe (around 2500-2800 years of age) was found in a collection of Egyptian papyri housed in Austria and called for one part rock salt, one part dried iris flowers, two parts mint and grains of crushed peppercorns.

Early Americans brushed with tooth powders containing harsh abrasives including baking soda, eggshells, alum, ground seashells, bone and even gunpowder. To freshen breath, powders included scented spices such as cinnamon or musk. A Connecticut dentist, Dr Washington Sheffield invented the first modern toothpaste around 1873, trialling it on his patients before forming the Sheffield Dentifrice Co. and manufacturing small batches of the paste in a facility behind his New London home in the 1880s. In 1892, he introduced the collapsible toothpaste tube, modelled on artist's paint tubes, which were quickly mimicked by Colgate and other major companies and with which we are familiar with today.

Left page Toothbrushes today exhibit a range of different colours, designs and bristle arrangements. **Above left** A battery-powered electric toothbrush. **Above center** A photo from 1899 showing the use of a toothbrush. **Above right** Dental technician Elizabeth Jarry shows an Iraqi girl, tooth-brushing techniques during a 13th Corps Support Command Medical Civil Action Project during Operation Iraqi Freedom. **Inset** 'Indexo' finger toothbrush, New York, United States, 1901-19.

Great Inventions We Take For Granted

breakfast cereal

The choice of millions of Americans every morning, ready-to-eat breakfast cereals have a surprisingly short history despite cereal grains being farmed and harvested for millennia.

Ancient North Americans had pounded and cooked corn into grits and Europeans had pioneered eating oats, but 20th Century America was, without question, the home of the ready-to-eat, cold breakfast cereal…even if the story began in the century before.

In late 19th Century America there was a fashion for the wealthy to flock to sanitariums or health retreats where clean air, healthy food and regular exercise were promoted as cures for all ills. Whilst working at the Danville Sanitarium in New York in 1863, Dr James Caleb Jackson developed his concentrated grain cakes called Granula. These required overnight soaking to be even remotely palatable.

The Kellogg brothers, John Harvey and Will Keith, both worked at another sanitarium in Battle Creek, Michigan. Firm Seventh Day Adventists, the pair experimented with a range of ways to create healthy foods with John Kellogg naming his first cereal product, Granula, which he changed to Granola after James Jackson sued. One batch of wheat the brothers had boiled was abandoned and left unattended, it dried out. Ever thrifty, the brothers pressed the wheat with rollers to try and make flat dough but the pressed grains turned into flakes. When they toasted the flakes and served them with cold milk at the sanitarium, they were surprised at the positive response they received. John Kellogg received a US patent in 1896 for his flaked breakfast cereal which he sold under the name Granose Flakes. The Kelloggs applied a similar technique to boiled, dried, rolled and toasted corn with Will buying out his brother and in 1906 opening the Battle Creek Toasted Corn Flake Company which later became Kellogg's.

Texan Charles William Post was a visitor to the Battle Creek Sanitarium in the early 1890s. Impressed by his time there, he opened his own health resort in Battle Creek and marketed a more convenient, bite-sized version of James Jackson's Granula which he branded as Grape-nuts as well as his own version of corn flakes. Cereal consumption rocketed and companies like Post, Nabisco

No.12 breakfast cereal

> "At the time I little realized the extent to which the food business might develop in Battle Creek."
>
> Will Keith Kellogg

and General Mills competed with Kelloggs and others. New cereals hit the marketplace with dizzying frequency. Many didn't stand the test of the time, but others such as Rice Krispies (invented by Kelloggs in 1928), Shredded Wheat (introduced by Nabisco in 1930) and Lucky Charms (invented by General Mills in 1961) remain firm favourites in America's kitchens each morning. The country's number one cereal in 2017 was the result of General Mills experimenting with 500 different formulae and flavours before Cheerioats debuted in 1941 and became Cheerios four years later.

Americans buy over 2.7 billion boxes of breakfast cereal each year. In 2017, 139.1 million boxes of Cheerios were sold in the US making it the single most popular cereal.

Left page (top) Shredded Wheat was introduced in 1930 by Nabisco. **Left page (far left)** More land is given over to wheat than any other single crop. **Left page (center)** Lucky Charms cereal invented by General Mills in 1961. **Left page (right)** A scoop of muesli. **Above** A bowl of Kelloggs Cornflakes. **Above (top right)** Advertising for Kellogg's Toasted Corn Flakes. Featuring 'The Sweetheart of the Corn', 1910s. **Above** The consumer has a wide variety of cereals to choose from today.

33

Great Inventions We Take For Granted

A shift of ironing, and you can step out into the world, confident of your creases, and with your wrinkle worries behind you.

the iron

However much one considers ironing drudgery today, it is a process far improved from that experienced in the past.

The ancient Chinese used pans filled with hot coals placed on cloth to remove wrinkles whilst unearthed Viking burial sites have discovered large pieces of glass used to smooth linen cloth. Glass stones shaped with handles and called by a variety of names including slickers or sleekstones were used in medieval Europe and before blacksmiths started creating flatirons – flat-based pieces of iron, heated in a fire and then run over clothing whilst still hot. Soapstone and terracotta-based hot irons were also used, first in Italy, France and the Netherlands.

Emerging through the Industrial Revolution, heavy sad irons featured stout iron bases and handles which were heated on stoves or in fireplaces. They could weigh 10 pounds or more and needed considerable stamina to wield them over long periods, punctuated by reheating or frequent scalding from the hot handles. In 1871, Ohio-born Mary Florence Potts developed a new type of heated iron. It had two pointed ends, was hollow to save weight and its most innovative feature was fitted to the base's top – a wooden handle that could be detached as the iron was heated and then reattached to lift away from the heat source and iron with.
Marketed as Mrs. Potts' Removable Handle Iron, sets included three bases and a single handle for 65 cents. Potts' innovation spawned dozens of imitators.

In June 1882, Henry W. Seely from New York City patented his electric iron – the first practical design, but it wouldn't be until 1905 that an electric iron enjoyed any significant degree of commercial success - the Hotpoint iron, developed by electricity meter reader, Earl H. Richardson of Ontario, California. Electric irons, with one or more elements making use of resistance to turn electricity into heat, offered greater convenience and marathon ironing sessions if needs be without having to reheat the iron. As increasing number of homes gained a mains electricity supply, so sales of electric appliances like irons increased. The New York Times in 1925 stated that 2.5 million electric irons were sold the previous year.

No.13 the iron

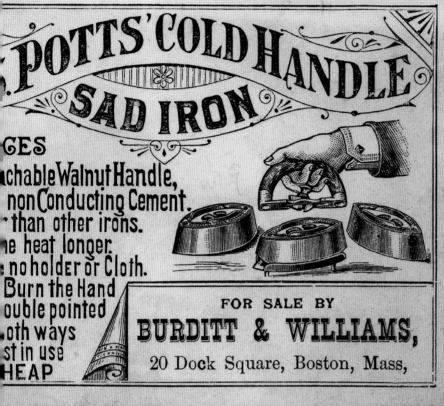

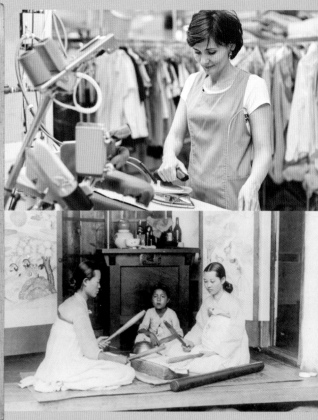

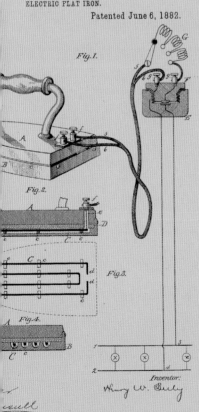

"I find it soothing to take something wrinkled and make it smooth. It feels anticipatory. It's what I do before a celebration. And nobody bothers me when I'm ironing."

Alexandra Stoddard, author and interior designer

Left page Three antique irons, the leftmost, an early electric model. **Above (left)** Mrs Potts cold wooden handle sad iron kits manufactured 1876–1950 by American Machine. **Above (top right)** A final iron for clothing at a dry cleaners. **Above (right)**Korean women, perform traditional Dadeumi ironing beating and rolling clothes using cylindrical bats (c.1910s). **Below** A typical modern day electric iron. **Left** Patent of the first electric flat iron in 1882.

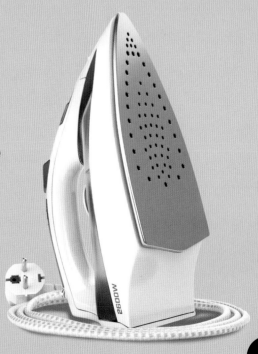

35

Great Inventions We Take For Granted

aluminum foil

Aluminum foil's excellent thermal insulation and unique barrier properties block out light, moisture and odors, making it a 'must-have' in the kitchen, at barbecues, for insulating wiring, electronics, and many other applications.

Above (from left to right) A roll of kitchen foil. Sealed foil pouches can hold dry or wet foodstuffs. Foiled wrapped baked potatoes emerge from an oven. Foil insulation is used between rafters to insulate a roof. Scored foil is used to cover chocolate confectionary. **Below** Chocolate Easter eggs are protected by a colorful foil wrapping.

No.14 aluminum foil

The shiny side of aluminum foil is 88 percent reflective, making it one of the best and most efficient insulators of solar heat, according to the Aluminum Association.

For much of the 19th Century, aluminum had been considered a precious metal. It was difficult to extract primarily from its chief ore, bauxite. The discovery of electrolytic reduction in 1886 and the industrial Hall–Héroult process enabled aluminum to be isolated and separated efficiently for the first time. At a stroke, aluminum became far cheaper to produce and new applications were quickly sought for the versatile metal.

Foils made of ultra-thin sheets of tin had been produced in the 19th Century and were used to wrap some foods although they could sometimes impart and leave a faint metallic taste. In 1910, Swiss engineer, Robert Victor Neher, applied for a patent to produce rolls of aluminum foil by a continuous rolling process as a replacement for tin. He opened a factory in Kreuzlingen, Switzerland the same year. The following year, a fellow Swiss company, Tobler, began using the foil to wrap its chocolate products including its famous triangular Toblerone bars.

Malleable and leaving no taste, aluminum foil was found to not absorb grease and oil whilst supremely effective at keeping heat in. By 1912, Maggi used foil as packaging for its stock cubes and soups with the first American product to be packaged in aluminum foil being Life Savers chewing gum and candy products in 1913. The same year came one of its first non-food uses when small bands of foil were wrapped around legs of racing pigeons in order to identify them in flight. In the United States, Richard Reynolds built a large corporation selling foil products to the catering and tobacco industries, began mining bauxite in Arkansas in 1940 and seven years later, introduced Reynolds Wrap Aluminum Foil to home consumers which proved a huge hit.

Since that time, aluminum foil has proven exceptionally versatile and is used not only in food and drink packaging (a thin layer of foil is often found in aseptic drinks cartons) but also as blister strip packs, wrappers, sachets and cartons for pharmaceuticals and other medical products. Foils are produced through double rolling – the rolling of two foil layers simultaneously – which results in the two different finishes, matt and polished with the matt created as the inner foil side during double rolling.

Approximately 7 billion aluminum foil containers are produced annually according to the Aluminum Association.

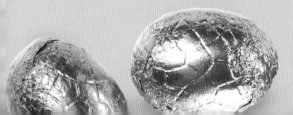

Great Inventions We Take For Granted

plastic wrap

Plastic wrap or cling film effectively began life as a greasy, dark green film used as a coating on World War II fighter planes and sprayed onto ships' exteriors to prevent corrosion from the salty sea air.

It had been discovered, according to legend, by a college student, working as a temporary cleaner at Dow Chemical, Ralph Wiley. Whilst at work in 1933, Wiley found a vial which, try as he might, he just could not scrub clean. He dubbed the strange cocktail of chemicals coating the container as Eonite, naming it after the indestructible substance found in the Little Orphan Annie comic books.

Dow researchers took note, developed the substance that Wiley had discovered which was polyvinylidene chloride (PVdC) and later refined and marketed it under the name, Saran, After initial use as an anti-corrosion coating and also in upholstery, it was refined and introduced to market as Saran Wrap in 1949 for commercial kitchen use. Four years later, it was sold to consumers for the first time.

Despite numerous other wraps now in existence, Saran remains the name most associated with the material that has proven an invaluable kitchen aid. PVdC's and its successor material, low density polyethylene (abbreviated to LDPE), possess long polymer chains which both stick to each other and repel liquid. Apart from its role as a handy food covering, and in use to protect brand new tattoos and in beauty treatments, people have discovered dozens of life hacks and other uses for the ubiquitous plastic wrap from a spectacles guard when painting and decorating to preserving bananas and slowing down their browning process - wrapping a banana in plastic film prevents the release of ethylene gas which speeds up the ripening process.

No.15 plastic wrap

Typical household cling film has a thickness of 0.0005 inches – about seven times less thick than the width of a human hair.

Cling film can keep cut vegetables fresh, prevent cross contamination in food storage areas and prevent leftovers from perishing, thus reducing food waste.

Great Inventions We Take For Granted

According to the Environmental Protection Agency, each baby in the United States goes through between 5,000 and 8,000 disposable diapers before they are toilet trained.

disposable diapers

No.16 disposable diapers

Disposable diapers today feature elasticized leg openings, either an elasticised or tab fastening waistline and a layered construction to take moisture away from the skin to keep the child dry. Typical diaper sizings range from N (for newborn) to 6 for infants weighing 35 pounds or over.

In 1946, Westport, Connecticut mother of two Marion Donovan was inundated with cotton diapers and soiled bed sheets caused by the washable diapers propensity to leak. The former assistant beauty editor of Vogue magazine vowed to do something about it, in the process making millions of parents' lives easier, and calling on her prior experience at the Indiana manufacturing plant run by her father and uncle where she had experimented and tinkered with devices and materials as a youngster.

Donovan designed and sewed her first breathable diaper covers from waterproof shower curtains and later, nylon material destined for parachutes, Unlike the rubber diapers available on the market at the time, her design caused little or no nappy rash nor pinched the baby's skin. She eventually obtained four patents for her design, which she called the Boater, and which eschewed safety pins in favour of metal and plastic snap poppers.

Despite hawking her innovative product around many leading manufacturers, Donovan struggled to gain any interest as she recounted to Barbara Walters in 1975, "I went to all the big names that you can think of, and they said: 'We don't need it. No woman has asked us for that. We don't need it at all,'….So, I went into manufacturing myself."

Donovan's Boaters went on sale in Saks Fifth Avenue in 1949 and proved a success. Within two years she had sold rights to the Keko Corporation of Kankakee, Illinois for $1 million. She then worked hard to produce a fully disposable diaper that wicked moisture away from the baby's skin but again, struggled to gain interest from large commercial manufacturers in the early 1950s. Later that decade, a Proctor & Gamble engineer, Victor Mills, who had worked on products as varied as Ivory Soap, Duncan Hines cake mixes and Pringles chips, was asked by the company to come up with innovative ways to use waste paper. Mills developed the idea of using shredded absorbent paper in a diaper that could be discarded after use. After tests using a Betsy Wetsy doll that mimicked urination and trialling his invention on his own grandchildren, Mills' brainwave eventually went on sale in 1961 under the brand name, Pampers®. Today more than nine out of ten babies wear disposable diapers and in 2012, Pampers® became P&G's first brand to record annual sales of $10 billion.

Great Inventions We Take For Granted

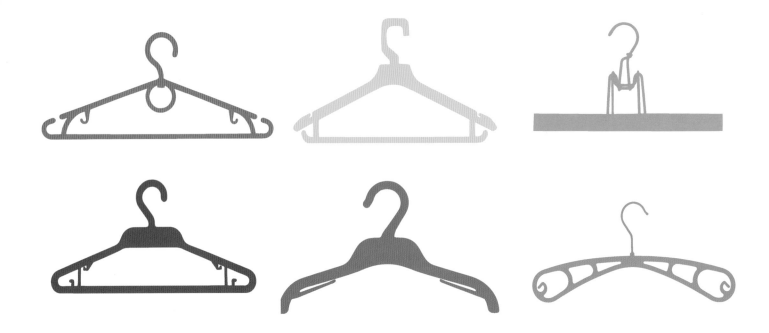

coat hangers

As humble and unheralded as household objects come, hangers have attracted more innovation than one might realize, with more than 200 patents in the United States alone.

No.17 Coat Hangers

More than eight billion plastic and wire hangers are sold each year, enough to fill over 49,000 semi trucks and form a convoy 464 miles long, according to Green Progress. Only 15% are ever recycled.

Cityscape Coat Hangers, each carved with the skyline of London, Milan, Tokyo, New York or Paris, first went on sale in 2004 at a price of $460 per hanger.

There was no one Eureka moment with coat and clothes hangers, no single point which marks the divide between their absence and their now current ubiquity. British monarch, Queen Victoria is said to have been gifted a set of coat hangers made from wood for her wedding day in 1840 whilst twenty-nine years later O. A. North from New London, Connecticut patented a wide oval-shaped coat hook with helped clothes keep their shape when hung up.

The Timberlake Wire and Novelty Company based in Jackson, Michigan has a legitimate claim to the title of first wire coat hanger manufacturer, applying for a patent in January 1904 and granted two years later, and possibly obtained from one of the company's employees, Albert J. Parkhouse. This hanger featured the familiar hook and stretched triangle shape but embellished with multiple coils or wire along its shoulders to act as supports. Dozens of US patents were granted in the first three decades of the 20th Century for rival designs. In 1932, Schuyler C. Hulett mounted cardboard tubes on the wire sections which came into contact with the clothing to prevent the wire marking the clothes; a hanger design still used extensively in dry cleaning today.

Alternative Uses

The humble hanger has seen service far and beyond keeping your clothes hung and crease-free. Wire coat hangers especially have been repurposed for many tasks – from homespun television aerials and a source of welding wire to a device to open a car door or clear a congested drain. One common life hack today is to bend and re-shape a hanger to make an impromptu stand for an iPad or other computer tablet. Coat hangers have also been life-savers, literally. When Paula Dixon suffered a collapsed lung on a 1995 British Airways flight between London and Hong Kong, two doctors, Angus Wallace and Tom Wong improvised at 35,000 feet by unfurling a wire coat hanger, sterilising it in brandy and using it as a trocar to stiffen a catheter as the lung was drained and the patient saved.

Above Today, hangers come in a bewildering variety of materials and designs from broad polycarbonate suit hangers or wooden or fabric padded dress hangers to multiple pants hangers and folding models for travel.

Great Inventions We Take For Granted

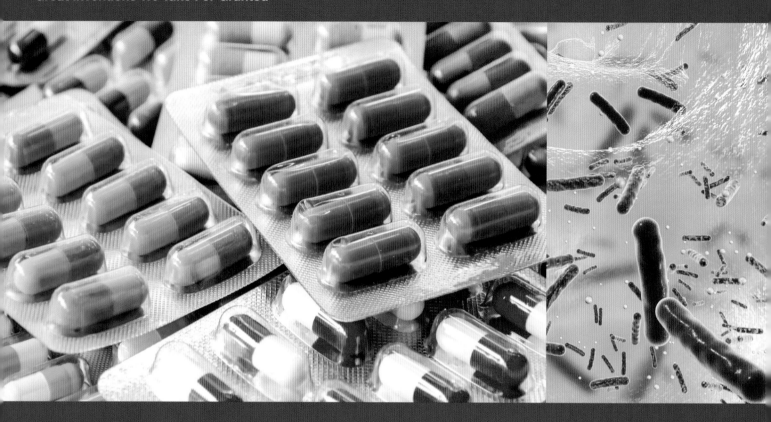

antibiotics

The story of antibiotics and how they're prescribed to tackle infections traditionally starts in 1928 with Scottish physician and microbiologist, Alexander Fleming and his serendipitous discovery of penicillin.

In truth, antibiotics have a far longer history, having been used for dozens of centuries earlier to treat infections long before people understood the science behind both the problem and solution. Ancient Egyptian medical papyruses, for instance, prescribed the use of mouldy bread pressed on wounds and scalp infections more than 3,000 years ago.

Some understanding of infections and ways of tackling them emerged in the late 19th Century leading to chemists seeking out, "magic bullets" that could selectively target and kill the microorganisms responsible for a painful, harmful or potentially lethal infection without harming the rest of the body. The laboratory run by German scientist Paul Ehrlich, a close friend of germ theory advocate, Robert Koch, had success in 1907 by synthesizing the antibiotic compound, arsphenamine. Two years, Sahachiro Hata, a Japanese bacteriologist working in Ehrlich's lab, discovered that arsphenamine could tackle syphilis. It was marketed under the name Salvarsan by Hoescht AG the following year.

In 1928, Scottish bacteriologist Alexander Fleming returned to his London laboratory from a holiday in the south of England and examined one of his unwashed culture plates containing Staphylococcus bacteria. He noticed a circle of mold surrounded by a halo where the bacteria had been wiped out and unable to grow back. Fleming identified the mold as Penicillium notatum and grew a fresh strain to repeat the process. He found it prevented Staphylococcus growth even when diluted 800 times and tackled other harmful bacteria as well.

Fleming named the mold's anti-bacterial component, penicillin but isolating it in quantities large enough to act as an effective treatment took more than a decade. Pioneering work by a team in Oxford, England led by Australian scientist, Howard Florey and Ernst Chain used homespun kit including hospital bedpans and cow milking equipment to produce enough penicillin to run medical trials on mice and then, in 1941, humans.

By then, in the midst of the Second World War, the demand for treating war wounds and combating diseases such as pneumonia,

No.18 antibiotics

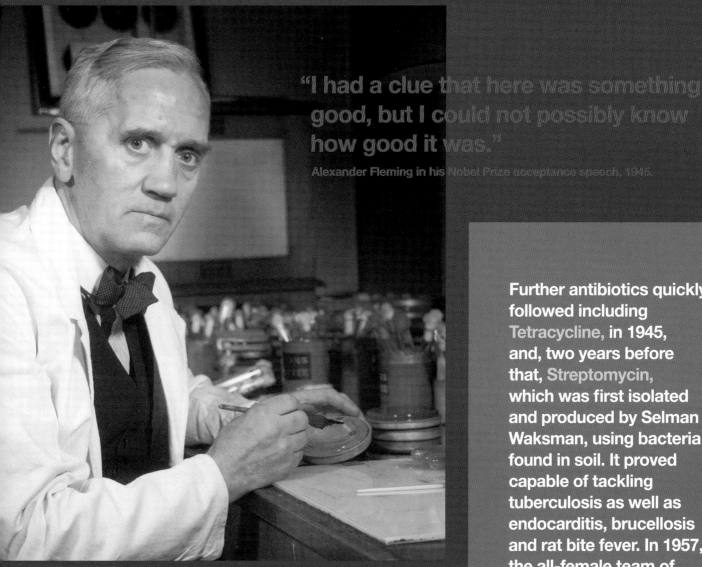

"I had a clue that here was something good, but I could not possibly know how good it was."
Alexander Fleming in his Nobel Prize acceptance speech, 1945.

scarlet fever, gangrene and diphtheria was great. It took enormous organisation and will of multiple agencies and pharmaceutical companies in the United States to find ways of producing penicillin on an industrial scale yet by 1945, the year that Florey, Chain and Fleming were jointly awarded the 1945 Nobel Prize for Medicine, some 560 billion units of penicillin were being produced per month, saving thousands of lives in wartime alone.

Left page (left) Foil-wrapped antibiotics capsules.
Left page (right) A 3D illustration of antibiotic resistant bacteria inside a biofilm.
Above Professor Alexander Fleming in his laboratory at St Mary's, Paddington, London.
Right Skeletal formula of a Streptomycin tuberculosis antibiotic (aminoglycoside class) molecule.

Further antibiotics quickly followed including Tetracycline, in 1945, and, two years before that, Streptomycin, which was first isolated and produced by Selman Waksman, using bacteria found in soil. It proved capable of tackling tuberculosis as well as endocarditis, brucellosis and rat bite fever. In 1957, the all-female team of Elizabeth Lee Hazen and Rachel Fuller Brown patented the antibiotic, nystatin, use in fighting fungal infections.

45

Great Inventions We Take For Granted

Whether it's to ease the pain of a sudden headache or to reduce the effects of over-indulgence at a bar, party or dinner the night before, millions of people reach for the aspirin packet every day.

aspirin

Around 100 billion tablets are taken every year yet, 120 years ago, this drug was simply not available to Americans or anyone else. So how did this absolute essential in everyone's medicine cabinet come about?

The ancient Egyptians had discovered the pain-killing and fever-reducing effects of a substance called salicylic acid at least 3,500 years ago. The Ebers papyrus, an ancient medical text dating from this time, mentions willow as an anti-inflammatory for aches and pains. The acid is found in certain plants including willow tree bark and the herb meadowsweet, also known as spirea. Pioneering ancient Greek physician, Hippocrates prescribed cups of willow leaf tea to both ease fevers and to aid women during childbirth, whilst in North America, the Blackfoot, Iroquois and Cherokee tribes also used willow tree bark for pain relief.

In 1828, Joseph Buchner, a professor of pharmacy at Munich University, managed to extract the active pain-relieving ingredient in willow. The result was small, yellow needle-shaped crystals with a distinctly bitter taste which Buchner named salicin after Salix alba – the white willow. A decade later, an Italian chemist working at the Sorbonne in Paris, Raffaele Piria split salicin into a sugar and a second component which he converted via oxidation and other processes into a colourless crystal which he named salicylic acid. This provided the pain relief but at a cost. The acid had an extremely unpleasant taste and many patients found that it caused vomiting and stomach upsets.

Charles Frédéric Gerhardt, a Frenchman working out of his own chemistry school in Paris, managed to buffer salicylic acid for the first time in 1853, reducing its damaging effects on the stomach, but he lost interest and his work was not taken up by others. It would not be until the end of the 19th Century that chemists working in Germany, notably Felix Hoffman at German pharmaceuticals company, Friedrich Bayer & Co, buffered the salicylic acid commercially to produce a stomach-kind pain relieving agent called acetylsalicylic acid (ASA).

The new drug went on sale in powdered form in 1899 under Bayer's own brand name of Aspirin. The A in the name came from acetyl chloride, "spir" represented the spiraea (meadowsweet) plant that the salicylic acid

No.19 aspirin

Pioneering ancient Greek physician, Hippocrates prescribed cups of willow leaf tea to both ease fevers and to aid women during childbirth

was derived from and the "in" was added to round off the name as a popular suffix for medicines. A year later, Bayer introduced soluble Aspirin tablets – the first time that a medicine had been sold in this form. Some 44,000 tons of tablets soluble or swallowed whole are now produced and consumed each year by people for headaches, joint pain, fevers and other health problems.

Left page Regular strength aspirin tablets tend to offer a 325 milligram (0.011464 oz) dose.
Above Willow bark and Spiraea (top) a genus of plants native to the temperate northern hemisphere are natural sources of salicylic acid.
Right German apothecary aspirin from Friedr. Bayer & Co.

> "America is the country where you can buy a lifetime supply of aspirin for one dollar and use it up in two weeks."
>
> John Barrymore, US screen and stage actor.

Great Inventions We Take For Granted

> "Access to communication in its broadest sense is access to knowledge and that is of vital importance to us."
> Louis Braille

braille

This ingeniously simple system of raised dots has transformed the lives of many blind and visually impaired people.

French boy, Louis Braille (1809-52) lost the sight in his right eye at the age of three after an accident in his father's saddlery workshop. Two years later, his left eye failed, leaving him totally blind. Braille was sent to the Royal Institute for Blind Youth. The school's founder, Valentin Haüy, had invented his own books for the visually-impaired by embossing large letters on pages which could be read by students tracing out the letters' outline with their fingertips. Tracing each letter out, though was slow and sometimes confusing, and with large letters, the books contained only the scantest of information. The books were also phenomenally time-consuming and expensive to create.

Whilst at the school, Braille learned about a complex system of night writing involving raised dots and dashes designed for the French army by Charles Barbier. Braille began experimenting to come up with his own, simpler, system. In 1824, at the tender age of 15, he invented an elegant solution using a grid made up of two columns of up to three raised dots. Braille's six dot system meant that 63 different combinations were possible – enough for the letters of the alphabet, basic punctuation and the numbers 0-9. This mini grid meant that each pattern representing a character could be touched and registered by a single fingertip, helping to make tactile reading far faster.

In 1829, Braille published a modest thirty-two-page book describing his system, Method of Writing Language, Plain Chant and Music, by Means of Raised Points for the Use of Blind Persons. He became a teaching assistant and then a teacher at the Royal Institute for Blind Youth and in 1837, the school published their first Braille book, A Brief History of France only three copies of which are known to have survived (one preserved in the American Foundation for the Blind's Rare Book Collection).

Today, Braille has proven vital in making life easier for the visually-impaired. Thousands of Braille books exist in libraries, some stretching to many volumes as Braille takes more space to display the words. J.K.

No.20 braille

Rowling's Harry Potter and the Goblet of Fire, for example, consists of 10 Braille volumes whilst the New American Bible extends to 45 volumes. Braille symbols appear on some store goods, medicines, household appliances and on bank teller machines. Braille can also be written as well as read using a machine called a braillewriter to produce the patterns of raised dots.

Braigo
Braille printers emboss paper with Braille characters to allow hard copies of files sent over a computer network. They tended to be large, complex and expensive until 2014, when 13-year-old Shubhan Banerjee from California, developed a small, low-cost Braille printer called Braigo using components mostly derived from a LEGO Mindstorms kit. Originally, designed as a science fair project, Banerjee received great publicity and funding from a variety of sources to form a company to develop Braigo and other assistive technologies.

Braille Benefits
"Studies have found that Braille is an efficient and effective reading medium with students demonstrating a reading speed exceeding 200 words per minute," stated the National Federation of the Blind in 2009. There are also professional benefits to learning braille. A survey conducted by Louisiana Tech University – the Transforming Braille Project Charter –found that people with a visual impairment who learn to read through braille have a much higher chance of securing a job.

Left page (left) Reading a braille book using fingertips. **Left page (center)** Braille symbols on the keypad of an automated teller machine (ATM). **Left page (right)** Braille symbols on the page. The letter A is the only letter of the alphabet represented by just one raised dot in the braille code. **This page (Top)** A braille printer outputs the pattern of raised dots to provide a valuable hard copy that visually-impaired people can read. **Above (left)** A refreshable braille computer keyboard and display features plastic or metal pins which rise to form the characters displayed on screen. **Above (right)** Braille symbols denote the floor numbers on an elevator's control panel.

"Braille's invention was as marvellous as any fairy tale. Only six dots! Yet when he touched a blank sheet of paper, lo! it became alive with words that sparkled in the darkness of the blind! Only six dots! Yet he made them vibrate with harmonies that charmed away their lonely hours!"

Braille, the Magic Wand of the Blind essay by Helen Keller c. 1924.

Great Inventions We Take For Granted

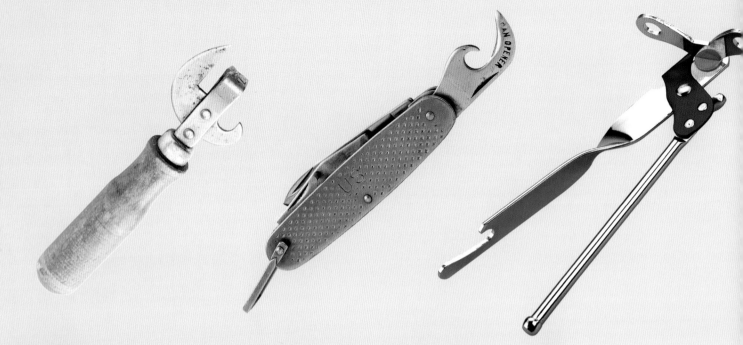

the can and can opener

Storing and preserving food occupied the minds of scientists, military men and politicians from the age of exploration onwards.

With armies, navies and explorers venturing further afield than ever before, portable stores of fresh foods preserved to not spoil, were a priority. Peoples had long extended the shelf life of food by preserving them using one of a number of techniques including drying, salting and pickling.

At the urging of French leader, Napoleon, the French Ministry of the Interior, offered a large prize of 12,000 French francs for anyone who could preserve meat, fruit and vegetables in a safe, convenient form that an army on the march could employ. The winner was Nicolas Appert, a candy maker from Massy, south of Paris. He devised a system of sealing food in glass jars and bottles placed in boiling water as a primitive method of sterilisation to kill off harmful microorganisms. Appert's bottling system was made public in 1810, prompting one French newspaper to praise him – "Appert has found a way to fix the seasons," but the glass containers were heavy, fragile and sometimes exploded due to pressure inside the jar.

Across the English Channel, the following year, English merchant Peter Durand was granted a patent by King George III for a similar sealed system but using iron cans covered in a thin layer of tin to act as a non-corrosive coating. Although the name on the patent is Durand, the brains behind the innovation was a Frenchman, Philippe Henri de Girard. Thwarted by bureaucratic red tape in his home country, Girard had visited London and demonstrated his canning system to members of the prestigious Royal Society.

Durand had no desire to develop Girard's invention – he was only in it for the money, selling the patent for one thousand pounds sterling to a pair of men, one of whom, Bryan Donkin, opened the world's first canning factory in London in 1813. Similar canneries followed throughout Europe and the United States initially making stout, hefty cans from wrought iron which later on in the century were replaced by lighter, thinner and cheaper steel cans coated in tin which we are familiar with today.

50

No.21 the can and can opener

Left page (left) A claw can opener punctures the can lid and then cuts around. **(center)** A penknife can opening blade **(right)** A butterfly or bunker style can opener - a type found in millions of kitchens around the world. **Above (left)** The four 50th Anniversary "Art Of Soup" Campbell's Tomato Soup cans featuring a facsimile autograph by, portrait of, and quote from, Andy Warhol. The commemorative cans were released in 2012. **Above (center)** A sardine can uses a twist and roll key to open its lid **Above (right)** Can lids and bases have concentric circle surfaces to aid stacking **Below** A can's opener's cutting wheel does its job whilst the outer, serrated feed wheel propels the can round as its key is turned.

"The advantages of my improvement over all other instruments for this purpose consist in the smoothness and rapidity of the cut, as well as the ease with which it is worked, as a child may use it without difficulty, or risk."

Extract from Ezra J. Warner's patent for a can opener, 1858.

Can Opener

For almost half a century, there were cans but no dedicated can openers. Users were instead advised to wield heavy hammers and chisels to break through the metal and prise open the can to access its contents. Whilst a primitive claw can opener was invented in England in 1855, the first US can opener was invented three years later by Ezra J. Warner from Waterbury, Connecticut. Warner's opener featured a guard which stopped the sickled blade from entering too deeply after it had punctured the can. A second curved blade would then cut around the top of the can using a sawing motion.

Warner's can opener was adopted by the US Army during the Civil War (1861-65) but not in the domestic arena where many viewed cans as something of a novelty. Grocery stores often stocked an opener to pre-open tins for customers before they took them home. New designs of can opener followed such as the first rotating opener that cut around the edge of the can, created by another Connecticut resident, William Worcester Lyman in 1870. In 1925, a company in San Francisco, the Star Can Opener Company, produced its Star opener. With its serrated wheel and circular cutting blade, it is the forerunner of the classic handheld opener people deploy to this very day.

Great Inventions We Take For Granted

Cotton swabs, or buds, reside in every household's medicine cabinet and many cosmetics boxes and drawers.

the cotton swab

Their many uses include as an applicator of cosmetics and ointments, make-up and nail polish removers and a cleaning tool for everything from electrical contacts to toy models. Just don't push them down your ear canals and harm yourself.

Leo Gerstenzang was born in Warsaw, Poland but in 1912, moved to Chicago and served during World War I. After becoming an American citizen in 1919 and relocating to New York City two years later, Gerstenzang observed his wife trying to clean their baby's ears by wrapping small pieces of cotton wool around a toothpick to create a small enough swab. He experimented for two years before perfecting a machine which could automatically wrap soft cotton wool around both ends of a wooden stick. Gerstenzang opted for smooth birch wood sticks in an attempt to avoid splinters.

In 1925, he launched his buds under the name Baby Gays but later changed the name to Q-tips with the, "Q" standing for "quality". They sold for 25 cents for 60 swabs, each of which had been dipped in a boric acid solution before they left the factory to ensure they were sterile. It became part of an entire line of baby and infant care products including Q-Talc and Q-Soaps. The wooden shaft remained until 1958 when the Q-Tip corporation bought out a British company that made paper sticks for candy and confectionary. Cotton swabs with paper sticks prevailed for many years until companies switched to plastic sticks, mostly Polyethylene terephthalate (PET), but recent environmental concerns over plastics pollution has resulted in a number of cotton swab manufacturers switching back to paper, wood or bamboo.

Above Bamboo shafted cotton swabs. **Right page (above left)** A cotton swab is gently eased into the ear canal to remove. excess wax.**Right page (above center)** Black coloured cotton swabs sold in Japan. **Right** Wild cotton grows in large balls on a farm in Adana, Turkey.

No.22 the cotton swab

"The problem is that this effort to eliminate earwax is only creating further issues because the earwax is just getting pushed down and impacted further into the ear canal."

Seth Schawarz, American Academy of Otolaryngology

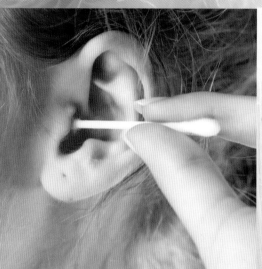

Many cotton swabs sold in Japan have their cotton coloured black, not white, after the Heiwa Medic Co. introduced the buds in the early 2000s. Customers believe they work better as anything removed by the swab is more clearly viewed on its black background

Great Inventions We Take For Granted

> "I couldn't get men to do the things I wanted in my way until they had tried and failed on their own, and that was costly for me. They knew I knew nothing, academically, about mechanics, and they insisted on having their own way with my invention until they convinced themselves my way was the better, no matter how I had arrived at it."
>
> Josephine Cochran

the dishwasher

Inventions have removed much of the drudgery from food preparation whilst the dishwasher comes to the rescue in the aftermath of a meal.

The first patent for a mechanical dishwasher dates back to 1850, and was granted to Joel Houghton from Ogden in New York State. His simple device was operated by turning a crank by hand to splash water on dishes and did not prove particularly effective or commercially successful. Other inventors paraded dish cleaning devices but they all lacked the force to remove stubborn stains from crockery and cutlery until a wealthy socialite grew tired of her servants chipping and cracking her vintage china crockery.

Josephine Garis Cochran's grandfather, John Fitch had run the first steamboat service in the United States, developing his own method of propulsion via a steam engine to power banks of oars that first propelled the Perseverance along in 1787. His granddaughter showed a similar strain of ingenuity in seeking a solution to the washing up problem. Working out of her home in Shelbyville, Illinois, Cochran fashioned wire compartments made-to-measure to expressly fit her own crockery. These compartments sat inside a wheel that lay flat and enclosed inside a copper boiler. A motor or hand turned crank turned the wheel whilst jets sprayed hot water onto the dishes and plates. She received her first of a number of patents in December 1886.

Cochran eventually formed the Garis-Cochran Manufacturing Company and, in what was very much a man's world of business at the time, had to battle to sell her machines. The dishwasher received a welcome boost in interest and sales when exhibited at the 1893 World's Columbian Exposition in Chicago and sold to large hotels, restaurants and canteens. Smaller, domestic models invented by British engineer, William Howard Livens in 1924 and an electric powered dishwasher developed by German appliance company, Miele in 1929, didn't catch on but included many of the features, from front-loading trays to rotating spray bars found in today's machines.

Dishwashers today use a pump to draw in water from a mains supply where it is heated by an electric element to within 86–140°F depending on the wash program selected. The detergent's release is timed to coincide with the hot water's exit under pressure from spray bars to dislodge debris and stains and

No.23 the dishwasher

Above (left) A front-loading domestic dishwasher contains control buttons and a small display indicating wash programme selection. **Below (left)** The bottom spray bar inside a dishwasher will spin and fire high pressure water to clean and rinse dirty homeware. **Above (center)** An early domestic dishwasher built by Willard and Forrest Walker of the Walker Brothers Company in Syracuse. **Above (right)** Advertisement in an 1896 issue of *McClure's* for The Faultless Quaker Dishwasher.

clean the load followed by rinse cycles aided by the use of a rinse aid which contains surfactants that lower surface tension so the rinse water runs off objects without streaking them.

Improvements in efficiency in the past 25 years and more microprocessor controlled washing programs in the latest dishwashing appliances, means that a high quality, fully-loaded dishwasher can use both less water and less energy than washing the equivalent amount of homeware by hand in the sink.

What Not To Dishwash

Certain items are best left for the sink including wooden utensils which may warp, copper cookware and aluminum pans (unless they're specifically labelled "dishwasher-proof") as their finish will dull over time, also cutlery with bone, plastic or wood inlay handles. A reaction can occur between silver cutlery and stainless steel, so these should not be washed in the same load, either.

Great Inventions We Take For Granted

For millennia, people have subjected pieces of bread to concentrated heat to turn it from cold, soft and pliable to hot, crispy and brittle toast.

the electric toaster

"Bread has been a staple part of our diet for 6,000 years, but toasting is relatively new and it's interesting that the process hasn't changed that much in 100 years. The 1926 Toastmaster looks pretty similar to the toasters we have in our kitchens today."

Spokesperson for bread manufacturers, Kingsmill.

No.24 the electric toaster

Toast Art

New Zealand artist, Maurice Bennett began experimenting with toast mosaics as a child. It culminated in his depiction of da Vinci's Mona Lisa painting, made from more than 6,000 slices of toast which was displayed at the K11 shopping mall in Hong Kong.

Above (left) An instruction manual cover for an early Toastmaster electric toaster. **Above (top center)** A modern four slot toaster offers convenient control of toast temperature and browning for both pairs of slots. **Above (below center)** A side opening antique toaster, c1930s.

The word actually comes from the Latin "tostum", meaning, "to burn or scorch." and was achieved by placing bread on a hot stone in a fire or holding it over the flames in a wire cage or on the end of a toasting fork. The concentrated heat produces the Maillard reaction between reducing sugars and amino acids in bread which browns the bread and gives toast its distinctive crunch and flavour. The same reaction is what gives roasted coffee its distinctive dark, smoky flavour and burgers their char-grilled taste. Named after the French chemist, Louis-Camille Maillard who described the reaction in 1912, the Maillard reaction occurs best in bread when heated in the approximate temperature range of 250-350 degrees Fahrenheit.

In the 1890s, electricity was making its way into the public realm mostly through public works projects such as street and train lighting. A Scottish engineer, Alan Alexander MacMasters was working on lighting for the Northern Line in London's Underground railway when he discovered the heat emitted by a cheap iron light filament. Working in the London laboratory of a fellow engineer, Evelyn Crompton, he invented an electric toaster in 1893 with four pieces of metal acting as heating elements fitted to a ceramic base. Crompton's company marketed the resultant simple toaster as the Eclipse. A more efficient and effective wiring for a heating element emerged in 1905 when a Pontiac, Illinois-born engineer in his twenties, Albert L. Marsh developed an alloy of nickel and chromium which was initially marketed as Chromel but became better known as nichrome.

The first commercially successful toaster, using nichrome wires, was developed by Frank Shailor at General Electric in 1909. Shailor's D-12 device consisted of a vertical cage in which a bread slice was placed and a single heating element toasted the bread. The D-12 only toasted one side at a time and had to be flipped by hand. It came with a plain or decorated base made of porcelain, cost $3.00 ($4.00 for the decorated model) and made its way into a number of diners and restaurants.

During World War I, a mechanic working in Stillwater, Minnesota called Charles Perkins Strite observed how much of the toast he saw served at a cafeteria was burned. He developed a system involving elements either side of the bread slot so that that both sides of the bread were toasted at the same time. Strite also fitted a mechanical timer and springs which would stop heating the toast with the springs ejecting the toast up and away from the heating elements for easily handling by the user. Strite patented his pop-up toaster invention in 1921 and five years later was selling his toasters under what would become the famous Toastmaster brand. The very first model, the A-1, held a single slice; two and four-slice models were developed later.

The arrival of pre-sliced bread in the late 1920s and early 1930s gave toaster sales an unparalleled boost. As mass production increased and manufacturing costs per unit reduced, toasters became affordable to all. Bimetallic sensors became common in toasters which helped regulate the temperature of the heating elements whilst push button ejection, heat control and the ability to toast different bakery products at different rates became a feature of microprocessor-controlled smart toasters in the 21st Century.

Great Inventions We Take For Granted

the safety pin

Great Inventions We Take For Granted

No.25 the safety pin

> "For all their low-tech utilitarian nature, safety pins are clearly a successful product design - they have been around since before the first modern Olympics."
>
> Andy Dixon, editor, Runners World

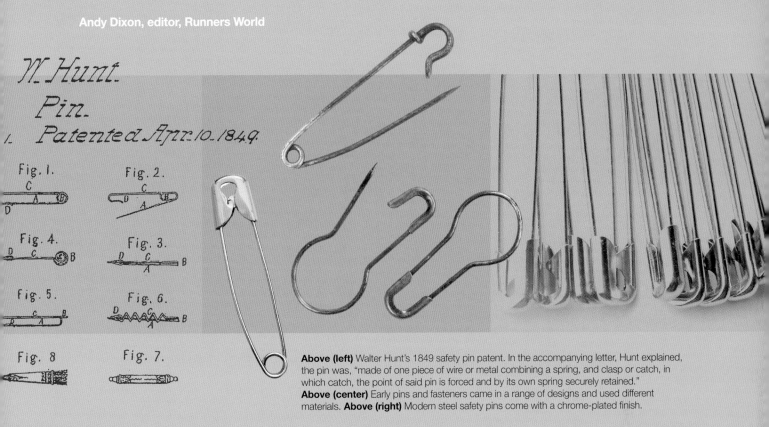

Above (left) Walter Hunt's 1849 safety pin patent. In the accompanying letter, Hunt explained, the pin was, "made of one piece of wire or metal combining a spring, and clasp or catch, in which catch, the point of said pin is forced and by its own spring securely retained." **Above (center)** Early pins and fasteners came in a range of designs and used different materials. **Above (right)** Modern steel safety pins come with a chrome-plated finish.

A pin whose sharp point is enclosed in a covering that also acts as a catch, may seem the most mundane of items, but it is actually a remarkable piece of design. Safety pins are fashioned inexpensively from a single piece of wire to create a spring, hinge, point and, sometimes, clasp. Their uses are many and varied and extend beyond sewing and quilting - from fastening cloth diapers to affixing bandages and athlete's numbers to their vests prior to competition.

Sharpened points of bone, wood or metal have been used since time immemorial to affix clothing together and in place. According to the Roman historian Herodotus, women in ancient Athens more than 2,300 years ago, were barred from using long, sharp pins to hold their cloaks in place after they had been used as a murder weapon, killing an Athenian soldier. Pins with a clasp and called fibula, meaning broach, stem back to before classical Ancient Greece to the times of the Mycenaean civilisation which inhabited the Mediterranean island of Crete. The earliest known example is dated to approximately 1100BCE.

Some three millennia later, Walter Hunt was struggling to make a profit as a mechanic and serial inventor living in New York State. Born in 1796, Hunt was prolific in his ideas back lacked the connections, foresight and business savvy to fully capitalise on his innovative ideas. Amongst his many inventions he failed to truly capitalise upon were a street-sweeping machine, a new design of tree-felling saw, an attachment for boats that enabled them to travel through icy rivers and most notably, a lockstitch sewing machine that might have had a profound impact if he had developed it further. Instead, by 1849, Hunt found himself struggling financially and anxious to repay a fifteen dollar debt.

According to legend, Hunt was fretting over his finances and fiddling with a piece or wire in his workshop when he discovered the spring tension in a single piece of wire coiled in its middle for the two ends to be held together under tension via a catch or clasp. Locking and unlocking pins could be made from a single piece of wire and not feature separate parts such as bases and hinges. Hunt sketched out the idea and obtained a patent in April 1849 before promptly selling the patent for $400 to a New York-based firm, W.R. Grace and Company who went on to manufacture millions of these small, metal household helpers.

Great Inventions We Take For Granted

The idea of a mechanical device to stitch garments together was an unmistakeable early sign of the impending Industrial Revolution...

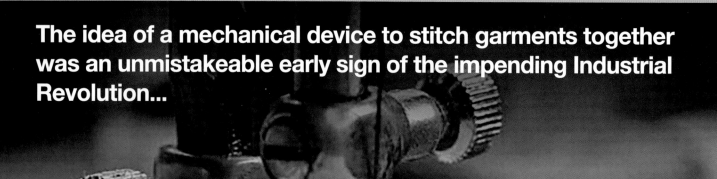

the sewing machine

The first-ever patent for a sewing machine to aid the craft of sewing was lodged in 1755 by Charles Wiesenthal, a German-born engineer living in England. But it was in 1790 that the English inventor Thomas Saint designed the first sewing machine, though it needed steady improvements over the next few decades before it became a practical proposition.

With the arrival of the 19th Century came a rapidly expanding urban population as a consequence of increasing industrialisation as well as a burgeoning middle class in many countries. Both helped create a growing demand for clothing. Whilst machines like the spinning jenny, waterframe and power loom had all helped mechanise textiles production, there was a desperate need for machines to help speed up the transformation of manufactured fabric into saleable garments.

Many engineers and entrepreneurs, sensing a potential fortune if they could perfect sewing by machine attempted to build devices with only a modicum of success. Englishman John Fisher invented a lace-making machine in 1844 which had many of the attributes of a sewing machine whilst French tailor Barthélemy Thimonnier's 1829 wooden model was the first to successfully perform straight, chain stitching. Thimonnier earned a contract to construct French army uniforms but his factory was later burned to the ground, allegedly by luddites who feared his machines would cause mass employment.

Above (left) A rare Davis domestic vertical feed (walking foot) sewing machine on a cast iron base, produced around 1890. **Above (right)** Closeup of a sewing machine needle.

No.26 the sewing machine

By 1900, sewing machines were making not only clothing, but awnings, tents, sails, cloth bags, book bindings and book manufacture, flags and banners, pocketbooks, trunks, valises, saddlery, harnesses, mattresses, umbrellas and linens.

Walter Hunt would gain fame as the man who invented the modern day safety pin and 'gave it away' selling his 1849 patent for the bargain basement price of just $400. Some fifteen years before, Hunt invented the first sewing machine with two spools of thread that performed lock stitching – where the needle thread passes through the material and interlocks with the second, bobbin, thread with the threads meeting at the center of the seam. Hunt didn't develop the machine to market. .

In April 1845, Massachusetts farmer, Elias Howe demonstrated his practical sewing machine powered by a hand turned wheel which advanced the cloth and a needle with an eye at its point to perform lockstitching. Howe's machine proved more than five times as rapid than the swiftest human seamstresses and received a U.S. patent the following year. He struggled to sell in the United States and travelled across the Atlantic to strike a deal with William Thomas, a British manufacturer of corsets and shoes.

On returning to the U.S. Howe discovered others building machines in breach of his patent, most notably Isaac Merritt Singer with whom he engaged in a five year legal struggle. Singer argued that he had made substantial improvements to Howe's model, including a way of controlling the thread and combining a vertical needle with a horizontal sewing surface, but the courts upheld Howe and Singer had to pay to licence, initially $1.15 for every machine he produced.

Later legal wrangles would see Singer, Howe and others join forces to form a patents pool called the Sewing Machine Combination, where all manufacturers in the group were able to benefit from each other's innovations for a small fee whilst others paid a prohibitive charge to licence the technology. Sewing machines advanced and sales boomed, so much that by 1913, Singer was, by capitalization, the seventh largest company in the world.

Top A threaded sewing machine ready to sew, showing two threads, one below from the bobbin, one above from the threaded needle on the foot. **Above** Drawing of the first patented lockstitch sewing machine, invented by Elias Howe in 1845 and patented in 1846.
Below A Jones Family machine from around 1935.

Great Inventions We Take For Granted

> "No one wanted to ask for it by name. It was so taboo that you couldn't even talk about the product."
>
> Dave Praeger, author of Poop Culture

toilet paper

It's quite extraordinary to think of the variety of different materials employed before the invention of toilet paper.

People tended to use what came to hand from large leaves and palm fronds to old rags, pottery shards in ancient Greece (a painful prospect) and flat sticks, known as chügi, in Japan which were drawn from left to right over the soiled area. Wealthy French noblemen and women in the 16th Century are said to have used pieces of lace, silk or hemp. In ancient Rome, toilets were mostly public and communal without cubicles or dividing walls between seats. Wiping was achieved using a communal sponge tied to an end of a stick and placed into a bucket of heavily salted water after use.

Early American settlers and farmers were not adverse to wiping with a corn cob but as printing and publishing expanded greatly in the United States and with it, the rise of newspapers and the mail order catalogue, it became common practice to fit a hook or nail to the inside of the lavatory door to hang squares of newspaper or old catalogue pages.

As the inventors of paper over 1,800 years ago, it is not surprising that the ancient Chinese would be first to produce toilet paper. Historians remain unclear as to precisely when this came about there are references to toilet paper as far back as the 6th Century where a court official and artist called Yan Zhitui wrote, "Paper on which there are quotations or commentaries from Five Classics or the names of sages, I dare not use for toilet purposes."

Court records from 1391CE show how China's Bureau of Imperial Supplies was producing 720,000 sheets for the ruler of Nanjing and his family. These sheets were large at approximately three feet by two feet.

One has to jump forward over 450 years to New York City to find the inventor of modern toilet paper. Massachusetts-born Joseph C. Gayetty came up with the idea of flat sheets of 'medicated' paper (500 sheets to a box) scented and watermarked with his name as a medical aid to those people suffering from haemorrhoids. These first went on sale in 1857 and were licenced for sale by other companies until as late as 1920. Between those times, the first toilet paper fitted to a roll were introduced by two different inventors in 1879, Walter Alcock in Britain and Clarence and E. Irvin Scott in the United States. The two Scott brothers, owners of the Scott Paper Company, began selling toilet rolls under the Waldorf brand but kept their name off or as small on the packaging as possible due to the embarrassment associated with the subject in 19th Century America.

No.27 toilet paper

The average American four-person household uses over 100 pounds of toilet paper a year, according to the National Resources Defense Council. Whilst the US comprises just over 4 percent of the world's population, it accounts for over 20 percent of global toilet paper consumption.

Left page Toilet rolls can vary in size from 160 sheet economy rolls to 1000 sheet jumbo rolls. **Above (left)** Leaves, shards of ceramics and corn cobs were all used for wiping in the past. **Above** A roll of toilet paper. Traditionally, sheets measured 4.5 inches square although some smaller sizes are produced. **Above (right)** Toilet rolls move on a production line. Some factories can produce two million rolls per day.

Companies proved less coy in the 20th Century as perforated squares of paper on the roll and double ply paper (introduced first by the St. Andrews Paper Mill in 1942) became common. Advertising campaigns featuring beautiful women, first depicted by the Hogan Paper Company's Charmin brand from 1928 onwards, four years before they introduced the iconic four roll pack, and later bears, dogs and other animals, pushed the product to the forefront of people's minds. By the 1970s, Americans were so enamoured with their soft toilet rolls that a jokey monologue by Tonight Show host, Johnny Carson in December 1973 about a fake toilet paper shortage prompted a real life one as people panic bought supplies and wiped supermarket shelves clean.

According to National Geographic, around 27,000 trees each day are cut down to supply the global demand for toilet rolls.

Great Inventions We Take For Granted

> "He was not a guy who was inspired by business. He was inspired by science. If you need to know how something works, sometimes you just need to know."
>
> Fraser Cameron, president and CEO of Velcro (2014-18) on George de Mestral.

velcro

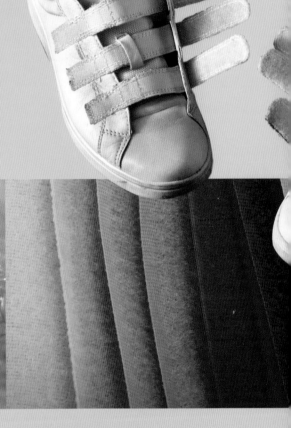

George de Mestral was a curious soul. Born in 1907 in Saint-Saphorin-sur-Morges, 10 miles away from the Swiss city of Lausanne, he had earned his first patent, for a new type of model aircraft, at the tender age of 12. He trained and worked as an electrical engineer and often walked for enjoyment in the foothills of the Alps mountain range. In the early 1940s, after taking such a walk, he returned home and removed the legions of sticky burrs of burdock plants that had affixed themselves to both his trousers and his Irish Pointer dog's fur. Curious as to their sticking power, de Mestral placed burrs under a microscope and saw their hundreds of tiny hooks which caught on to the tiny loops in his clothing, and the dog's fur. This gave him the idea for a touch fastening device for clothing.

De Mestral spent more than a decade perfecting an artificial version of the burrs. The practical difficulties involved in mass manufacturing the tiny hooks at a specific angle to latch onto loops occupied him whilst the technical challenge of producing a material with some 300 tiny hooks in every square inch proved daunting. Yet in 1955, de Mestral had a complete practical system in place which he had patented. It consisted of pairs of nylon strips, one fitted with thousands of hooks which meshed with the other strip containing thousands of equally microscopic loops. De Mestral named it Velcro®, from the French words velours croché meaning "hooked velvet". Within four years, over 50 million metres were being produced each year.

Despite envisaging his invention as a replacement for buttons, zips and other clothing fasteners, Velcro® was not initially adopted by many clothing manufacturers but it was quickly used in hospitals for heart blood pressure cuffs and patient gowns. Velcro® received a big boost in sales after its adoption by NASA for use to counter microgravity in space and prevent objects floating away in weightless environments. NASA astronauts quickly got used to using pens, meal trays, toiletries and food bags all kept in place using strips of hooks and loops fasteners. The Scrabble board used on the International Space Station is Velcro-ed in place on a cabin interior and each tile is backed with the material to keep letters in place.

No.28 velcro

NASA astronauts quickly got used to using pens, meal trays, toiletries and food bags all kept in place using strips of hooks and loops fasteners.

In 1968, Puma became the first major shoe company to sell training shoes with Velcro® fasteners. By the 1980s, it was unusual not to see a child in lace-less trainers with three tell-tale Velcro® straps acting as fasteners. De Mestral's patent ended in 1978 leading to a boom in copycat materials and increasing uses – from securing limb braces and keeping car mats in place on floors to acting as an impromptu tablet or smartphone holder.

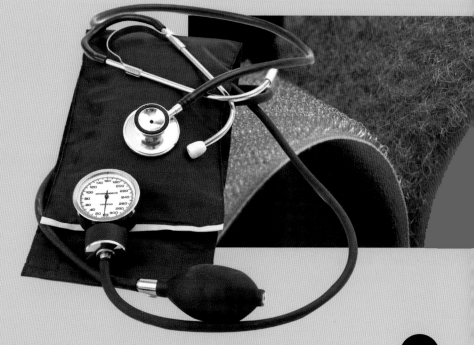

Left page (top) Velcro provides an easy fastening option for children's sneakers. **Left page (left)** The hooked burrs of a plant catch in loops in fibre and fur. **Left page (center)** Rows of nylon eyes on one Velcro strip catch hooks on another strip.
Left page (right) Rolls of velcro manufactured in an array of colours. **Above** NASA astronaut Greg Chamitoff, Expedition 17 flight engineer, ponders his next move as he plays a game of chess on a velcro board the International Space Station.
Right Velcro is used as a fastening on blood pressure monitor sleeves.

Great Inventions We Take For Granted

spectacles

The invention of glass lenses that could help correct eyesight defects have proven a boon from those suffering from farsightedness, nearsightedness, astigmatism and other vision impairments. Today, according to the Vision Council, 188.7 million Americans own spectacles to correct vision problems, all reliant on discoveries and inventions made in the past.

Ancient peoples discovered the magnifying power of pieces of smooth semi-precious stones such as beryl and quartz or curved glass if polished and held a short distance away from the object to be viewed. The earliest known magnifier of this type was unearthed at Ninevah near modern day Mosul, in Iraq, and is believed to date from 640BCE. Two Arab scholars, Ibn Sahl and Ibn al-Haytham (also known as Alhazen) produced important writings on the principles of optics between 984 and 1021CE. Al-Haytham's Book of Optics, in particular, explained how light travelled in straight lines and how the human eye collects and focus light to form an image on the retina at the back of the eye.

Al-Haytham's book proved influential when it was translated into Latin in Europe and in the 13th century was studied by Italian monks. Some of the first known spectacles are believed to have emerged from Italian cities such as Pisa, Florence and Venice around 1270-1290 with two convex lenses fitted into a simple frame made of wood or animal bone and held in front of the eyes as the frames had no arms. These glasses, referred to as rivet spectacles today due to the rivet at the center joining the two halves of the frame together, helped correct farsightedness. A 1352 painting by Tommaso da Modena in Treviso, Italy is the earliest depiction of spectacles with a pair perched on the nose of a cardinal.

With the invention of movable type printing in Europe in the 15th Century and, with it, increased literacy and availability of books and other written materials, the use of spectacles slowly grew. This century saw the first glasses with concave lenses that helped correct nearsightedness or myopia although the scientific principle underlying the innovation wasn't clearly explained until Johannes Kepler's 1604 work, Clarification of Ophthalmic Dioptrics. Spectacle frames finally got arms that balanced on top of the ears or pressed on the sides of the head in the 18th Century with London optician, Edward Scarlett, optician to King George II, often attributed as the inventor.

No.29 spectacles

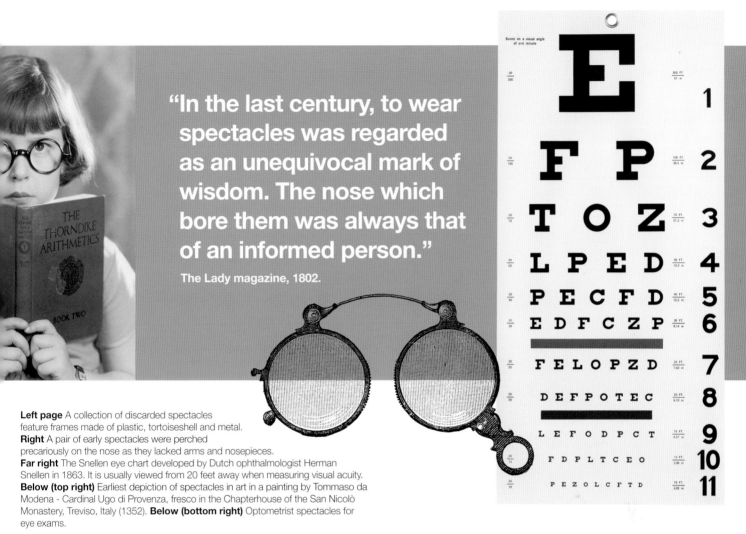

> "In the last century, to wear spectacles was regarded as an unequivocal mark of wisdom. The nose which bore them was always that of an informed person."
>
> The Lady magazine, 1802.

Left page A collection of discarded spectacles feature frames made of plastic, tortoiseshell and metal.
Right A pair of early spectacles were perched precariously on the nose as they lacked arms and nosepieces.
Far right The Snellen eye chart developed by Dutch ophthalmologist Herman Snellen in 1863. It is usually viewed from 20 feet away when measuring visual acuity.
Below (top right) Earliest depiction of spectacles in art in a painting by Tommaso da Modena - Cardinal Ugo di Provenza, fresco in the Chapterhouse of the San Nicolò Monastery, Treviso, Italy (1352). **Below (bottom right)** Optometrist spectacles for eye exams.

Two Into One

American statesman and polymath, Benjamin Franklin suffered poor eyesight from his youth and is believed to have worn spectacles from the 1730s to correct his myopia (nearsightedness). As he aged, Franklin found he also suffered from presbyopia – a decreasing ability to focus clearly on nearby objects – a common disorder as one grows older.

History is unsure on whether this inventor of the lightning conductor and the Franklin stove, either invented the first bifocal glasses himself or he instructed a lensmaker – possibly the optician, Samuel Pierce – to cut two sets of different strength lenses in half and place half of each lens in the frame – so that the bottom half corrected his near sight and the top half his long distance impairment. What we are certain is that Franklin popularised his double spectacles which in 1824 were dubbed "bifocals" by British inventor, John Isaac Hawkins, a man who invented trifocals with three different lenses as well as an upright piano design he sold to another American statesman, Thomas Jefferson, for $264 in 1800.

In 1959, Frenchman, Bernard Maitenaz invented the varifocal progressive lens for use in spectacles whilst working at Essilor. These lenses have a gradual change in strength from the top of the lens to the bottom with multiple focal points allowing you to see all distances and focus points through just one lens. In 2012 alone, 20 million progressive addition lenses were prescribed and dispensed in the United States.

Great Inventions We Take For Granted

super glue

No.30 super glue

> "Superglue, which, in addition to fascinating children, has actually saved lives as a means of sealing wounds."
>
> President Barack Obama, 2010.

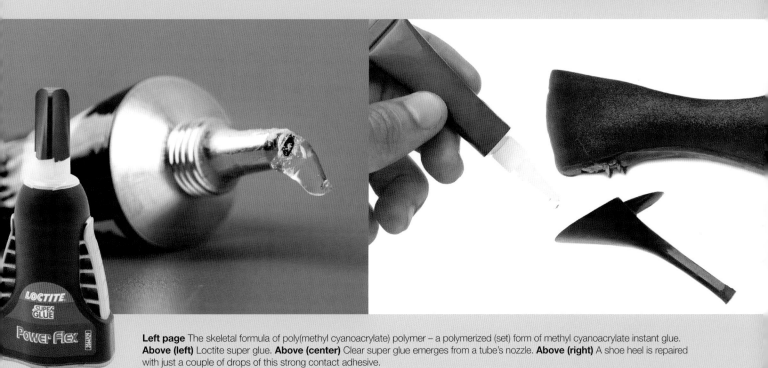

Left page The skeletal formula of poly(methyl cyanoacrylate) polymer – a polymerized (set) form of methyl cyanoacrylate instant glue. **Above (left)** Loctite super glue. **Above (center)** Clear super glue emerges from a tube's nozzle. **Above (right)** A shoe heel is repaired with just a couple of drops of this strong contact adhesive.

A surprisingly large number of inventions owe a degree of serendipity to their success. This was certainly the case with the world-famous glue which is found in most household's hardware drawers. In 1942, as part of the war effort, chemist Dr Harry Wesley Coover was working on the creation of crystal clear gun sights made of plastic. The former Hobart College and Cornell University chemist worked for Eastman Kodak and was experimenting with a form of cryanacrylic which whilst clear proved unsuited for the role as it stuck to anything it came into contact with.

Coover shelved the material until 1951 when he was working on polymers for heat-resistant cockpit canopies for military jet aircraft at Kodak's Kingsport plant in Tennessee. One of his technicians, Fred Joyner, found ethyl cyanoacrylate stuck two expensive refractometer prisms together firmly without the need for any heat or pressure. Re-discovering the substance, Coover suddenly appreciated its attributes and tested it out on various objects within the laboratory. Providing there was a small amount of moisture on the surfaces to be bonded, the objects stuck permanently each time.

Coover eventually registered a patent for the glue as an, "Alcohol-Catalysed Cyanoacrylate Adhesive Compositions/Superglue" and worked on refining the product for commercial production. Super glue went on sale first in 1958 under the name Eastman 910 before later being licenced to Loctite who dubbed it Loctite Quick set 404. During the 1970s, various manufacturers produced their own take on the fast-setting glue with the strong bond, using Coover's cryanacrylic formula.

Over the years, stunts demonstrating the extreme fixing power of the adhesive have been broadcast. Coover participated in one of the earliest and most memorable, on live TV when I've Got A Secret host, Garry Moore was raised into the air by a metal bar held together by just a single drop of glue. One of the most recent on the Russian TV programme, Chudo Tehniki (Wonders of Technology) in 2015 saw the show's producer, Dmitriy Demin, dangling upside down a hot air balloon at an altitude of 4,900 feet, held only by boots super-glued to a small wooden platform.

In 2010, Coover, who had been involved in obtaining 460 patents during his career, received the National Medal of Technology and Innovation from President Obama. He was most proud of the adhesive's lesser known but crucial role in medicine. During the Vietnam War, field surgeons found that cyanoacrylates sprayed over wounds incurred in battle could act as an emergency method of staunching blood flow. Today, newly-developed forms of cyanoacrylates are often used in surgery to close up incisions in conjunction with or in place of traditional sutures.

Great Inventions We Take For Granted

Today's HVAC and central heating systems are marvels of microprocessor-controlled efficiency and convenience, many controllable remotely via smart home phone apps. These systems build on the earlier inventions to provide key components of central heating including boilers and furnaces, radiators and thermostatic control of the system.

central heating

The ancient Romans were ingenious engineers, adapting others' inventions often on a grand scale. This was the case with the forerunners of modern central heating, the hypocaust, which historians believe the Romans adapted from an earlier Greek invention where channels or flues carried heated air up into structures such as the Temple of Artemis at Ephesus. The Romans' hypocausts were elaborate underfloor constructions consisting of a floor raised on brick columns called pilae. A large furnace burning wood and stoked by slaves generated heat in a hypocaust to warm air. This was channelled through the underground space made by the pilae to warm the floor, made of layers of ceramic tiles and concrete, from below. Channels or flues in the walls of the building enabled additional heat to rise up the building warming rooms, before it exited the building through openings called flues in the roof. Hot, rising air systems are thought to have persisted in China and ancient Korea, the latter's ondol system of heating from below via a stove or firebox, could still be found in some homes in Korea post World War II. The Roman-styled hypocaust, however, was largely lost with the fall of the Roman Empire in the late 5th Century CE.

In the late 18th and early 19th centuries, a number of steam engineers including Thomas Tredgold, legendary Scot, James Watt and the inventor of the first passenger-carrying steam locomotive, Richard Trevithick, produced early designs of steam boilers, but these were used predominantly for heating the large factories arising in the burgeoning textiles industry.
According to the English Heritage organization, the manufacturing of water heating boilers in large quantities began from the 1860s onwards with the first room heaters often comprising coils of pipe carrying hot water from the boiler with the pipes hidden behind decorative cases. True radiators began with the 1840s patents of American, Joseph Nason who developed a basic model in 1841 and the Russian aristocrat and entrepreneur, Franz San Galli who received a patent for his radiator design in 1855.

No.31 central heating

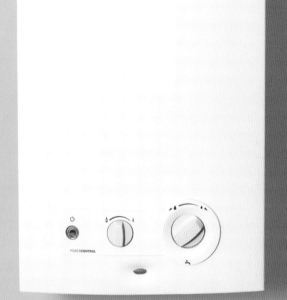

"People don't wear clothing to keep warm any more. One of the social norms is that people can go around in shirt sleeves at home or in the office. So central heating has changed the way people think about clothing."

Harry Charrington, UK architect.

Remains of hypocaust systems have been found throughout the widest extents of the Roman empire including England, Germany, Cyrene in North Africa and in Portugal such as this 1st Century CE hypocaust of the Cantaber Domus in the Roman settlement of Conimbriga.

Another key component of central heating systems emerged thirty years later with the invention of the first thermostat for regulating heating temperatures. Albert Butz, a Swiss-born engineer residing in Minneapolis, created the damper flapper, a movable door which regulated the amount of air entering the furnace; the door closed when warm to restrict air flow and lower heat output and rose when cold to force the furnace to burn more fiercely thus increasing heat output. Butz's patents were bought by Mark Honeywell in 1906 who would go on to develop the first programmable thermostat – the Jewell – with a built-in clock for users to time when their heating came on or off.

Left page (left) Boiler control panel offers adjustable regulation of hot water temperature as well as heating settings and timings. **Left page (center)** A radiator thermostatic valve offers straightforward adjustment by hand. **Left page (right)** A brass valve fitted to an ornate iron radiator. **Above (left)** Smart control panel household gas boiler.

Great Inventions We Take For Granted

the safety razor

Sharpened iron, copper, or bronze blades replaced shells and stones, and with them came the rise of the harrowing, straight, or 'cut throat' razor.

These took considerable skill to wield safely and effectively giving rise to thousands of barbers in cities to which men would flock for a shave.

As the world centre of cutlery making in the 18th Century, it is perhaps no surprise that the English city of Sheffield was where the first modern razor blades were fashioned, using a high grade steel first developed in 1740 by Joseph Huntsman.

Twenty-two years later, a French cutlery maker, Jean-Jacques Perret added a wooden sleeve or guard to a straight razor to limit the amount of blade that came into contact with the skin. This was the first known attempt at creating a safety razor that men could use themselves and others followed, featuring toothed or combed guards which could nestle between beard hairs.

In 1847, William Samuel Henson of Somerset, England turned matters around when he filed for a patent for a detachable comb tooth guard for a razor, "the cutting blade of which is at right angles with the handle, and resembles somewhat the form of a common hoe." The hoe-shaped razor's arrangement of a blade perpendicular to the handle made it far easier to grip and control than straight razors and was adopted in 1875 by the Kampfe brothers, from Saxony, Germany after they had moved to New York and opened a machine shop in Brooklyn. Their Star razor featured a blade just over one and a half inches long and proved popular, so much so that it spawned numerous imitators causing the Kampfes to advertise frequently that, "All Others Are Spurious", but between 1880 and 1901, over 80 safety razor patents were applied for or issued in the United States alone.

One of these applications was made by bottle cap salesman, King Camp Gillette who had noted how all razors needed seemingly constantly re-sharpening via a strop or a visit to a professional re-sharpener, and how a business model conducted around a disposable product could yield great profit. Gillette combined the hoe-shaped razor with a curved clamp that held a double-edged

No.32 the safety razor

King Camp Gillette c1907-1908.

"There is no other article for individual use so universally known or widely distributed. In my travels I have found the safety razor in the most northern town in Norway and in the heart of the Sahara Desert."

King Camp Gillette.

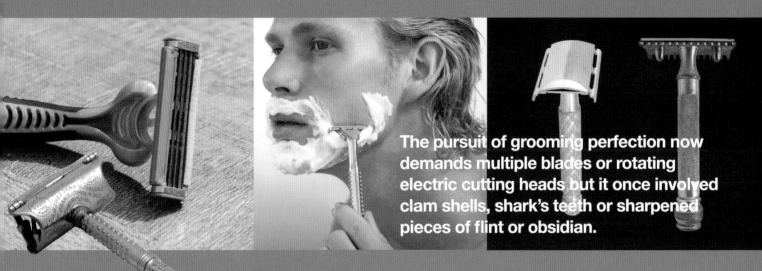

The pursuit of grooming perfection now demands multiple blades or rotating electric cutting heads but it once involved clam shells, shark's teeth or sharpened pieces of flint or obsidian.

disposable blade that once blunted was replaced with another. It took him and MIT-trained engineer William Nickerson a number of years to perfect the technique or stamping out blades from sheets of high-carbon steel, but by 1903 they had their first batch of razors ready for sale. Within three years, Gillette was selling more than 300,000 razors annually; the razors were initially sold at a loss, more than compensated for by the large profit made on replacement blades, over 70 million of which were sold in 1915.

Left page A steel razor blade.
Right An advert for the Star safety razor, published in Scientific American in December 1907.
Above (left to right) Early Gillette single-blade safety razors contrasted with the multi-blade models popular today. **Above inset** Vintage Kampfe Bros. Star Single Edge Safety Razor, made in the United States.

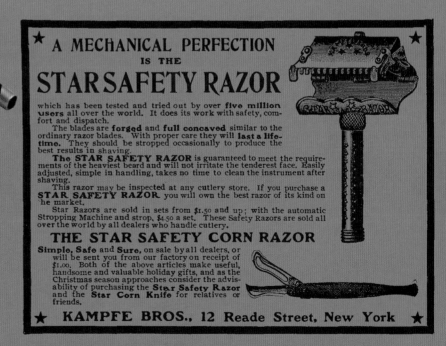

Great Inventions We Take For Granted

the cellular mobile phone

No.33 the cellular mobile phone

"We had no idea that in as little as 35 years more than half the people on Earth would have cellular telephones, and they give the phones away to people for nothing." – Martin Cooper

Communications on the move proved essential for military personnel and law enforcement officers whose vehicles from the 1940s onwards were often equipped with radiophones. Similar systems for commercial users were introduced into around 100 towns and cities by AT&T, starting with St Louis in 1946. The 'mobile' element was qualified as some 80 pounds of equipment was required to transmit and receive calls via VHF radio signals.

Researchers at Bell Labs and elsewhere began to experiment with the concept of a cellular phone network. The idea was to blanket the nation with a network of small hexagonal cells, each of which would contain a base station that ferried and received messages from mobile devices using radio frequencies, connecting the mobile signals to the main telecommunications network whilst the phones themselves would switch frequencies as the person using the device moved from one cell to another.

In the 1960s, the notion of personal wireless mobile communication with anyone anywhere in the world still felt fanciful and more like a gadget out of the realms of the newly broadcast Star Trek show and the communicators used by Captain James T. Kirk and the rest of the Enterprise's crew. Yet, it would be less than a decade before a handheld mobile phone was produced by Motorola and demonstrated by the head of its communication systems division, Martin Cooper. He cheekily called the office

Above (left) The original DynaTAC 8000X released by Motorola in 1983.
Above (right) Martin Cooper re-enacts his pioneering 1973 call with a DynaTAC prototype cellular phone for the media in 2007.

Great Inventions We Take For Granted

In the second quarter of 2017, the number of unique mobile subscribers globally passed the five billion mark according to GSMA Intelligence.

Above Users access their modern phones using touchscreen technology to connect via social media or text. **Right page (left)** A user scrolls through a webpage using their smartphone. **Right page (right)** IBM Smartphone with its charging cradle. **Below** The evolution of cellphone design has seen candy bar and clam shell models with small screens abd the focus on voice calls to large but slim smartphones used more for social media and running apps than making calls.

No.33 the cellular mobile phone

land line of his chief rival, Dr. Joel S. Engel at AT&T, in 1973, using the prototype DynaTAC phone. It measured 10 inches tall, weighed 2.4 pounds, more than half of its weight comprising batteries, and had a total call time of less than 30 minutes, taking some 10 hours to recharge. Cooper later joked how battery life wasn't an issue, "because you couldn't hold that phone up for that long!" Motorola would invest an estimated $100 million and the next decade on turning this concept device into a practical consumer good – the DynaTAC 8000X which was now 13 inches tall including its aerial, weighed 28 ounces and cost $3995 at its launch in 1983.

Motorola were quickly joined by other players in the cellular phone market but still notched several further notable firsts including the first flip-down phone, the MicroTAC in 1989 and seven years later, the light for its time 3.1 ounce StarTAC, one of the first phones to use rechargeable lithium-ion batteries and to come with a vibrate option. In between, IBM brought the Simon Personal Communicator to market. It was the first device to combine mobile telephony with personal digital assistant (PDA) features including a touchscreen, notepad, appointments scheduler and the ability to send and receive faxes and emails. As such, Simon is considered the forerunner of the modern day feature-packed smartphone, ironically used increasingly for data, social media and web access and less for the voice calls that cellular phone pioneers intended their devices' purpose.

Great Inventions We Take For Granted

soap

"I am constantly amazed when I talk to young people to learn how much they know about sex and how little about soap."

Billie Burke, US actress.

According to Ancient Roman legend, soap magically appeared through the forces of nature when rainfall landed on Mount Sapo where animals were frequently sacrificed, washing a mixture of wood ash from sacrificial fires and melted animal fats down into the River Tiber below. There, the soap-like mixture was said to be found useful for washing both people and their clothing.

Away from myths and legends, many ancient cultures stumbled on the dirt-removing properties of salts and fatty acids in the correct combination. Whilst a 'recipe' for manufacturing a cleaning agent appears in the ancient Egyptian Ebers Papyrus, dated to around 1550BCE, there is another civilization with an even earlier claim. Clay containers with a residue like soap from ancient Babylonia and dated to 2800BCE have been discovered whilst a 2200BCE clay tablet from the same civilization gives a formula for a cleaning substance's production involving water, an alkali and cassia oil.

True soaps are produced by the saponification or basic hydrolysis reaction of a fat or oil. Sodium carbonate or sodium hydroxide are commonly used to neutralize the fatty acid and convert it to a salt of a fatty acid. The resulting substance, soap, has long molecules with one hydrophobic end whilst the other end is hydrophyllic or water-loving. These work to pull dirt and grease away from skin or fabric into water to clean.

Post-Reformation Europe saw a rising demand for soap despite several monarchs taxing the substance brutally. Soap makers used a variety of fats and oils, from tallow derived from animal fats to olive oil – found in castile soap. Sodium carbonate (also known as soda ash) was in high demand. In 1791, French chemist Nicholas Lablanc patented a process that used common salt to make soda ash in large quantities and cheaply, slashing the cost of soap production at a stroke (the Solvay process in 1861 would further reduce costs). Thirty-two years later, another French chemist, Michel Eugene Chevreul published a detailed explanation of the chemistry of fats and fatty acids, the foundations of soap chemistry to this day.

No.34 soap

In 1891, New Yorkers were each given a free cake of Colgate soap as they waited their turn to try out the city's first public bath.

Left page Olive oil soaps from Marseille. **Above (left)** Pieces of soap on the assembly line at the plant for the production of soap. **Above (right)** Cut Handmade Soap. **Right** Soap making in 1875. **Below** Colorful soap in different colors from the Provence, France.

Dove bar of soap

A study by the U.S. Census and Simmons National Consumer Survey (NHCS) indicated that approximately 114 million Americans used Dove soap in 2017, making it the most popular brand.

Great Inventions We Take For Granted

"On a very local scale, a refrigerator is the center of the universe. On the inside is food essential to life, and on the outside of the door is a summary of the life events of any household."

Robert Fulghum, American author

the refridgerator

It's hard to believe that the humble household fridge is barely a century old, so much do we all rely on it for our supply of fresh, chilled foods and drinks.

In 1748, Scottish scientist William Cullen discovered that the evaporation of liquids involves the absorption of heat. This principle was adopted and exploited by a number of engineers with varying degrees of success. American inventor Jacob Perkins, living in London at the time, built the world's first working vapor-compression refrigeration system in 1834, using a closed cycle with ether as the refrigerant. Whilst working in Florida, physician John Gorrie produced a similar system in 1854. Both men's systems were commercial failures but others succeeded and industrial refrigeration became employed especially in the meat packing industry in the later 19th Century. By this time, a booming urban population in the United States and much of Europe meant that many people were far removed from farms and the food production process. Households often shopped daily for fear of spoiling food or stored fresh foods in shaded larder cupboards or in iceboxes which relied on regular deliveries of fresh ice.

In 1876, a German engineering professor, Carl von Linde, patented a process that liquefied gas that led to the first reliable industrial refrigerators using compressed ammonia. Frederick J Wolf obtained the rights to sell Linde's machines in the United States and his son, Fred W. Wolf junior invented the first domestic refrigerator using this process in 1913. His DOMELRE (Domestic Electric Refrigerator) was designed to sit on top of a kitchen's icebox and could be plugged into a light socket to obtain the power needed for its compressor and other parts. New fridge designs began emerging with alarming rapidity leading to companies like Kelvinator from 1914, Frigidaire from 1918 and Electrolux from 1923 all starting to ramp up production of refrigerators for the home. Many machines were large and in two parts needing installation of a separate compressor often located in a home's basement. They were also phenomenally expensive. A Kelvinator in 1922 cost $714 – equivalent to close to $11,000 today. General Electric's 'Monitor Top' refrigerator, introduced in 1927 and nicknamed after its circular compressor's resemblance to the ironclad warship USS Monitor, was smaller, a little cheaper and the first to be enclosed in all steel cabinet. Coinciding with more and more households gaining a mains electricity supply, it proved popular with GE producing over a million machines.

No.35 the refridgerator

The US FDA recommends a refrigerator be set to below 40 °F. Temperatures above may not inhibit the growth of potentially harmful bacteria.

How Fridges Work
Refrigerators are effectively heat pumps which move heat energy from cold to warm – the opposite direction to normal heat flow – in order to cool cold areas further. A closed system of pipes circulate refrigerant which is alternately compressed and expanded, changing state to draw heat away from inside the refrigerator. The refrigerant as a liquid under high pressure, flows through a one-way expansion valve into pipes running inside the refrigerator. The valve lowers the refrigerant's pressure suddenly, making it expand, cool and turn from liquid to gas via evaporation. As it does so, it cools the air inside the refrigerator. The cool air sinks, forcing warmer air upwards to also be cooled. The refrigerant then passes through a compressor which increases its pressure and temperature, turning it into a liquid again to begin another cycle.

Features and Failure
A number of features today taken as standard in refrigerator design were borne out of initial failure or only muted success. The butter compartment in the door where the air is a little warmer to keep the butter softer was first proposed by consultant Lurelle Guild, working for Servel in 1942. The company declined to include her suggestion but other manufacturers did adopt it. Similarly, Wolf's DOMELRE fridge failed to sell significantly but one feature, its ice tray, did catch on.

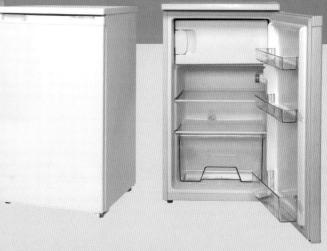

America led the world in adopting electric refrigerators. By 1950, 90 per cent of US urban homes had a fridge, compared to less than three per cent of British households.

Left page A 1960s fridge with cool water outlet.
This page A selection of free-stranding fridges from the past and present. Despite changes in external styling, greater internal space and more energy efficiency, modern fridges exhibit plenty of similarity to those of the past.

Great Inventions We Take For Granted

"Quick freezing was conceived, born, and nourished on a strange combination of ingenuity, stick-to-itiveness, sweat, and good luck."

Clarence Birdseye

frozen food

General Foods' opening range of 27 different frozen products bearing Birdseye's name was trialled in Springfield, Massachusetts starting on March 6th, 1930.

Whether stored in a small icebox built into a refrigerator or a capacious chest freezer, frozen food enables millions to enjoy a great variety of prepared meals and treats including fruit and vegetables out of season.

Peoples had preserved foods by salting, pickling or drying or storing in ice for centuries. The slow freezing of food in a snowdrift or in an icehouse often created large, disruptive ice crystals. When thawing, the crystals can damage cell membranes and dissolve emulsions causing foods' taste and texture to decrease markedly. Utilising new freezing technology, one of the first commercial freezing operations was founded in the 1860s in Sydney's Darling Harbour, but this Australian operation suffered from the same problem with slow freezing just below freezing point leading to largely tasteless products.

Living within the Arctic Circle, the native Inuit peoples froze much of their catch of fish and other marine life extremely rapidly in temperatures below -30 °F. Clarence Birdseye, a naturalist, supporting himself by working as a fur trader in Labrador, Canada in the 1910s, experienced the Inuits' work and tasted its results – the fast freezing resulted in far small ice crystal formation which disrupted cells less. Thawed Inuit trout and other fish tasted almost as good as freshly caught examples.

Birdseye returned to his native United States intent on replicating the Inuits' techniques but on an industrial scale. He trialled blanching and rapidly freezing vegetables straight after harvesting and continued his experimentation whilst working for the Clothel Refigerating Company in 1922. The following year he formed Birdseye Seafoods Inc. to produce and sell fish fillets frozen using chilled air. The company went bankrupt in 1924 but Birdseye was undeterred as he developed innovative new ways to flash freeze food by packing it into waxed cardboard cartons and freezing it rapidly between stainless steel plates to well below sub-zero temperatures. He gained patents for flash freezing technology first

No.36 frozen food

Left page Frozen mixed vegetables.
Above (right) A woman in the 1950s opens the small freezer compartment contained within her refrigerator.
Above (left) Supermarket freezers store frozen produce for customers. The frozen food industry is estimated to be worth over $56 billion annually. **Right** A selection of family favorite ice lollies and ice-creams.

using calcium chloride and then vaporising ammonia as a refrigerant and a double belt freezer system.

Birdseye sold his company and technology to Goldman Sachs Trading Corp. and Postum for $22 million in 1929 (the same year that Postum became General Foods) but continued to work for the company. He helped oversee a complete frozen foods infrastructure, including refrigerated boxcars and freezer cabinets in stores, to establish a farm-to-factory-to-foodstore system that has since revolutionised many people's daily diets.

"By modernizing the process of food preservation, Birdseye nationalized and then internationalized food distribution...facilitated urban living and helped to take people away from the farms...and greatly contributed to the development of industrial-scale agriculture."

Mark Kurlansky, author of Birdseye: The Adventures of a Curious Man

Great Inventions We Take For Granted

The last of the trinity of tableware to make it to mealtimes, forks exert a strange fascination. Maybe, it's due to their late arrival and the strong opinions they have provoked throughout history.

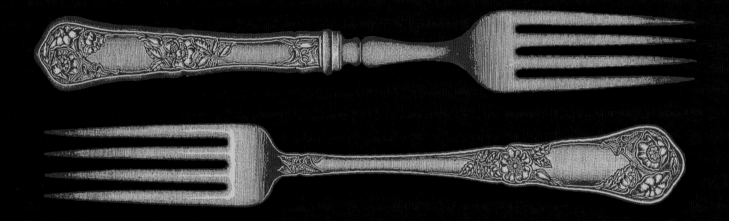

the dinner fork

"The two-pronged fork is used in northern Europe. The English are armed with steel tridents with ivory handles - three pronged forks - but in France, we have the four-pronged fork, the height of civilization." E. Briffault, writer in Paris a table, 1846.

The spoon and the knife have been part of mealtimes since prehistoric times. Ancient peoples developed knives from sharpened flint and other rocks, and later, metals, which were employed to cut tough meat and vegetables whilst simple spoons were formed from the use of large seashells or rounded chips of wood or tree bark. The Anglo-Saxon word for spoon, "spon" comes from a chip of wood.

Fork-shaped implements abound in ancient civilizations' mythology with the three-pronged trident of Poseidon prominent in ancient Greek myths and in Hindu scriptures and stories where the god, Shiva was said to wield such a weapon to keep evil, negative gods and spirits at bay. Whilst double and triple-pronged spears became used by ancient peoples for fishing, warfare and at the ancient Roman gladiatorial games, they rarely reached the dinner table. Some historians believe that a simple pointed implement may have been used much like a fork in ancient Babylonia and bone forks have been unearthed and attributed to the Qijia Culture of China that existed between 2200 and 1600 BCE, but for the first widespread use of the implement, we have to turn to Persia where a fork-like barjyn was used in the 9th Century CE and spread to the Middle East the following century.

One of the earliest recorded evidence of forks in Europe comes from 11th Century Venice and the wedding of a Byzantine noblewomen Maria Argyropoulina to the son of Pietro Orseolo II, doge (ruler) of this Mediterranean city-state. According to legend, Argyropoulina brought a set of gold forks with her as part of her dowry – an act that scandalised some Venetians. St. Peter Damian a monk and cardinal wrote, "God in his wisdom has provided man with natural forks – his fingers. Therefore it is an insult to Him to substitute artificial metallic forks for them when eating."

In the Middle Ages, most people ate using their fingers off large pieces of stale bread called trenchers which held meat, fish and vegetables and with the use of spoons or knives for anything else that couldn't be brought directly to the mouth. The modern use

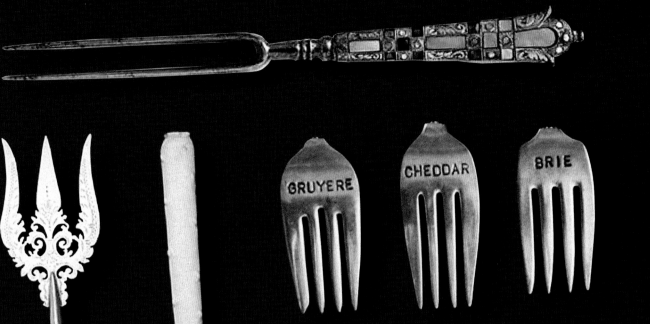

of forks developed out of their adoption and development in much of continental Europe with different nations including France, Italy and Poland still vying today to claim their forebears hastened the spread. What is certain that their use spawned a vast range of specialist forks for different courses or delicacies – from delicate oyster, crab and cocktail forks to large, robust beef forks and dessert forks with a widened edge to slice down through cakes and other sweet treats.

English travellers, merchants and gentlemen engaged in their 'grand tour' of the continent, picked up the habit of using a fork and slowly popularised its use back in Britain. One of the first was 16th and 17th century traveller, Thomas Coryat who remarked of his travels to Italy in 1611, "Forks are made of iron or steel: noblemen eat with silver forks. I have gone on using a fork even now that I am back in England. This has occasioned more than one joke." Forks were regarded as "foreign" and "unmanly" before their gradual mainstream adoption which occurred in America in the 18th and 19th Centuries.

Right page Classic Sheffield Steel Silverplate forks, 1890s.
This page (top) 11th Century Byzantine fork with jewelled gold inlay handle. **Above (right)** A set of six individual cheese forks, English, from the early Eighteenth Century. **Above (left)** A Victorian ivory-handled sardine fork. **Above (left)** An onyx-handled silver bread fork, Austrian Empire, 1860s

The first known example of something approaching a 'Spork' – a combined fork and spoon – was patented in 1874 by Samuel W. Francis in the United States and featured a knife blade fitted to one edge of the implement.

Great Inventions We Take For Granted

Phillumenists are collectors of matchbooks and matchbox covers, the word first being recorded in 1943. King Farouk I of Egypt (1920–1965) was a noted phillumenist who once took a dedicated flight from Egypt to Denmark purely to buy a single matchbox cover.

the safety match

We take the instant flame at our fingertips that a struck match can provide for granted. Yet, people struggled with flints and other troublesome devices before its development.

In 1826, John Walker, a pharmacist in the north of England developed the first friction match selling his cardboard stick matches, dipped in sulfur and tipped with a mixture of chlorate of potash and sulphide of antimony, in a small box accompanied by a small piece of sandpaper out of his store from 1827. Without a patent, Walker's product could be copied without retribution, a task performed by Samuel Jones who started selling his matches as, "Lucifers" two years later. Frenchman Charles Sauria used white phosphorous to produce popular matches that others adopted with Alonzo Dwight Phillips in 1836 the first to patent such a match in the United States.

White phosphorus is pyrophoric and had the dangerous tendency to spontaneously self-ignite. In addition, the terrible health problems that beset factory workers and their sustained handling and inhaling fumes from this toxic substance, led to public demand for a change. Swedish chemist, Jöns Jacob Berzelius and one of his students, Gustaf Erik Pasch figured out that separating out the friction match's inflammable chemicals and dividing them between the match head and the striking surface would create a safer match.

Pasch produced matches with heads coated in red phosphorous, a substance less flammable than white phosphorous and a specially-prepared striking surface covered in a mixture of phosphorous and powdered glass. When struck, the match would generate a small amount of white phosphorus which began the ignition process leading to the match burning. Pasch received a patent in 1844 and began manufacturing his new match but they proved expensive and it would be another pair of Swedes, the Lundström brothers, who would commercialise the safety match which by 1858, they were selling 12 million boxes of per year. Safety matches today use a mixture of non-toxic potassium chlorate (45-55% of the head's contents) as an oxidising agent along with glue, binder and additive compounds such as antimony (III) sulfide or sulfur to act as fuel.

Matchbooks
Cigar-loving Pennsylvania attorney Joshua Pusey is believed to have invented the matchbook, containing 50 matches and a

No.38 the safety match

Above (left) A match burst into flame as it is struck along the side of a matchbox. **Above (right)** A colorful collection of vintage matchboxes from Thailand. **Left** A matchbox full of Iga Bintang branded matches from Malaysia. **Right** Small matchboxes typically contain 40-45 matches, their sticks often made from white pine. **Below right** A collection of vintage match books.

striking surface on the inside of the cover, patenting his invention in 1894. Three years later he sold the rights for $4000 to the Diamond Match Company of Ohio. A young salesman at the company, Henry C. Traute, had a vision of the surfaces of the matchbook as mini billboards which smokers and other match users would stare at many times each day, and convinced companies to advertise on matchbooks despite their diminutive size – just two inches tall. In 1896, the Pabst brewery ordered thousands of matchbooks printed with their adverts inside and out, sparking the common practice of matchbook advertising.

work

Great Inventions We Take For Granted

Howard Fielding was a young boy when he became the first person in history to pop some bubble wrap. He was not the last. Many cannot resist bursting some of the air pockets when unwrapping a fragile object. The packaging material now even has a Bubble Wrap Appreciation Day which is celebrated in January on the last Monday of the month.

bubblewrap

Some inventions are conceived to perform a particular or fulfil one purpose only to find longevity and popularity performing a quite different role. Play-Doh modelling clay, for example, was invented in the 1930s by Noah McVicker at soap manufacturers, Kutol Products in Cincinnati, as a cleaner for wallpaper darkened and dirtied by soot from wood and coal fires. The same was the case with this famous packaging staple found in many homes.

Alfred Fielding was a young engineer who trained at Stevens Institute of Technology and after graduating was based in Hawthorne, New Jersey. Working in tandem with Swiss chemist Marc Chavannes, the pair fashioned plastic sheeting in a garage leaving air bubbles between the two layers to create what they thought would be a sure-fire winner – exciting new wallpaper for the start of the Space Age. It actually proved an inglorious failure and the pair realigned their invention as an insulating material for greenhouses, again without commercial success.

Showing remarkable perseverance, the pair persisted, this time repurposing their invention as a wrapping or packaging medium for delicate objects and forming the Sealed Air Corporation in 1960. Their timing proved fortuitous as IBM were looking to ship their new computer, the first to feature an all-transistor design and one of the first truly all-purpose machines, the 1401. IBM received an, at the time, unprecedented 5,200 orders in its first five weeks of release and Bubble Wrap proved an ideal way of protecting its vulnerable parts without the mucky fingers associated with balled newspaper. Fielding and Chavanne's company took off with the wrap used extensively in fresh food packaging, in industrial and commercial packaging and incorporated inside mailing envelopes.

Away from its role as the pre-eminent packaging material of delicate and valuable objects, Bubble Wrap continues to be used in unusual ways. Ironically, considering its commercial failure as a greenhouse insulator, some gardeners use it to line cold frames and insulate crops during cold snaps in the weather. It has been used as a homespun additional liner in sleeping bags and in January 2019, London Fashion Week Men's saw designer Craig Green's Fall 2019 collection include entire outfits fashioned out of coloured bubble wrap.

No.39 bubblewrap

"We get a lot of packages with Bubble Wrap in them, and it's really hard for me to part with this stuff."

Howard Fielding, son of Bubble Wrap's co-inventor, speaking 2018 to the North Jersey Record.

Bubblewrap comes in various sizes. It is typically made from resin using temperatures around 560 degrees Fahrenheit. A 12 inch square piece of this iconic material is housed in the Museum of Modern Art.

Great Inventions We Take For Granted

banknotes

Those folding bills in your wallet or back pocket have a long history.

China is believed to be the home of paper, where its making was first recorded by a Chinese court official called Ts'ai Lun around 104CE. Lighter in weight than papyrus and other writing materials, early paper was made by soaking a mixture of plant fibres such as tree bark and hemp along with shredded rags in water and then mashing the fibres into a wet pulp. This was strained and pressed onto a frame and then hung up or laid out to dry. The end result was a thin sheet of interlocked fibres that formed a flexible, easy to write on material.

A carefully-guarded secret in China for centuries, it is no surprise that the Chinese were the first to produce paper notes acting as a promise to pay the bearer a set sum of money, mostly held in copper coins or other valuable items. During the latter half of the Tang Dynasty (618-907CE) paper promises to pay were used by the government for the first time, sometimes to pay local merchants in far-flung parts of the Chinese empire. The Song Dynasty (960–1279 CE) introduced the first national paper banknotes, often referred to as Jiaozi, printed with red and black ink, stamped with seals to deter counterfeiting and with the year of issue clearly displayed.

It took centuries before banknotes reached Europe and the first major attempt ended in disaster. The Bank of Stockholm was formed by Johan Palmstruch in collaboration with the Swedish royal government in 1657. Four years later, Palmstruch introduced kreditivsedlar - credit notes with specific denominations as a substitute for Sweden's heavy copper coin currency. They were printed on watermarked paper and carried eight signatures as well as the bank's seal. Palmstruch unfortunately printed more notes than the bank had reserves to back up the circulation and the bank collapsed.

He was convicted, sentenced to death, but died in prison in 1671. Other European nations, however, followed suit with the Bank of England founded in 1694 producing handwritten bank notes and the first French Franc notes appearing in 1795.

Paper To Plastic
In 1988, Australia issued the first national banknotes printed on polymer rather than paper. Although more expensive to produce, these waterproof notes last around two and a half times longer than paper notes, stay cleaner and are considered more difficult to counterfeit. More than 20 other countries have followed suit

Making A Buck
According to the US Federal Reserve, a single one dollar bill costs 5.5 cents to produce on linen and cotton paper in two facilities – one in Fort Worth, Texas and the other in Washington D.C. Each bill lasts an

No.40 the banknote

In 1988, Australia issued the first national banknotes printed on polymer rather than paper.

average of 5.8 years before it is replaced. In 2017, there were an estimated 12.1 billion one dollar bills in circulation. The first $1 bill was issued in 1863 as a Legal Tender Note and carried a portrait of the Treasury Secretary, Salmon P. Chase. George Washington replaced Chase on the bill in 1869 and the current design stems from 1963.

Money from Money
Notaphilists collect bank notes, some of which have proven incredibly valuable. In 2014, an ultra-rare Grand Watermelon $1000 treasury note from 1890 went up for auction. The note which featured a portrait of Major General George Meade on the front, received its nickname from the design of its large zeroes on its reverse side which resembled watermelons. Only two are known to be in existence, prompting its sale for $3,290,000. It had last been sold in 1970 for a more modest sum of $11,000.

Left page A selection of colorful banknotes with contrasting designs from around the world. **Above (left)** Australian banknotes. The $100 design featuring singer, Dame Nellie Melba, was designed in 1996, the others in 2017 or 2018. **Above right (top)** A USA one dollar note. **Above** The obverse and reverse sides of an 1890 $1,000 Treasury note depicting U.S. Army soldier George Meade and featuring the watermelon-styled zeroes.

Great Inventions We Take For Granted

"The Macintosh uses an experimental pointing device called a 'mouse'. There is no evidence that people want to use these things."

Computer columnist John C. Dvorak, writing in The San Franscisco Examiner in 1984. He did return to the subject three years later, admitting he had been converted.

the computer mouse

Douglas Englebart was a computing visionary. After reading Vannevar Bush's futuristic article As We May Think, Engelbart devoted himself to improving computer-human interactions and innovations and the ability for computers to enable human workers to collaborate over computer networks. Taken for granted today, this was revolutionary stuff in the 1950s and early 1960s when Engelbart headed the Augmentation Research Centre at the Stanford Research Institute (SRI).

One of his chief projects was the oNLine System (NLS) which was the practical computing implementation of Engelbart's vision. It allowed people to share documents, pioneered word processing and featured a highly advanced interface for the time that allowed users to input information and commands via a cursor as well as copy, paste, change or move data on screen – a significant advance. Such a system required new input devices to control programs on screen. At the time, from its development and use with radar screens, the light pen was considered by many as the optimum input tool for controlling the computer system but Engelbart wasn't convinced.

His team experimented with a wide range of potential input and control devices. These include trackballs, a device operated by the user's knee under a workstation desk and the chorded keyboard where five bars were pressed in different combinations in a similar way to playing a piano chord, to perform tasks and commands. However, these fell short when pitted against Engelbart's idea of a wooden box with buttons. Built by Bill English, it featured two metal wheels at right angles to one another, as positional recorders. As the device was moved, the wheels turned and sent back electrical signals which were converted to move the position of the cursor on the computer screen. The team quickly found it outperformed the other devices for speed, ease of use and in minimising mistakes.

The team experimented with fitting as many as five buttons on the device before later settling on three for their later plastic moulded models. Engelbart patented the device as, "an X-Y position indicator for a display system" in 1967. It was demonstrated publicly for the first time

No.41 the computer mouse

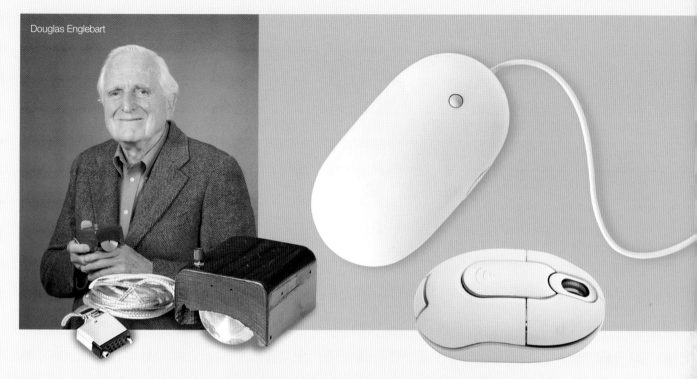

"Here's your design spec: Our mouse needs to be manufacturable for less than fifteen bucks. It needs to not fail for a couple of years, and I want to be able to use it on Formica and my blue jeans."

Steve Jobs of Apple to one of his designers, after visiting Xerox PARC and viewing their mouse in 1979 working on a Xerox Alto computer.

the following year at the 1968 Fall Joint Computer Conference in San Francisco. There, in front of 1,000 mostly stunned computer technologists, Engelbart and his team performed a 90 minute demonstration of cutting edge computing now remembered as, "the Mother of All Demos". The mouse featured heavily during the demonstration with Engelbart even telling his audience, "I don't know why we call it a mouse. It started that way, and we never did change it."

In 1972, Bill English, now at Xerox PARC, along with Jack S. Hawley, developed the first ball mouse replacing the original design's wheels with a rolling ball capable of monitoring movement in any direction, not just horizontally and vertically. The ball came into contact and moved two rollers that turned wheels.

The wheels' movement was converted into electrical signals giving the mouse's direction of movement and its speed. Hawley left Xerox PARC and began selling three button mice for around $415 whilst in 1982, Logitech sold its first mouse, the P4, for $299. Take up of early mice was slow until the mid-to-late 1980s when personal computers boomed, rising in performance and dropping in price. Apple made use of its own, single button design on LISA and Macintosh computer whilst in 1991,

Logitech released the first widely-available wireless mouse to use radio frequency transmission – the Cordless MouseMan. By 2008, the company shipped its one billionth mouse.

In 1996, Microsoft released the IntelliMouse Explorer, the first popular mouse to feature a scroll wheel as well as buttons. The rubberized plastic wheel located between the two standard buttons allowed users to scroll up and down windows and documents, amongst other functions. Later mice, especially those conceived for specialist computing such as engineering design and high level gaming, were developed with a series of programmable buttons and functions to enable precision control.

Left page (top) Xerox Alto mouse.
Left page (below) Apple cordless mouse.
Above right Douglas Englebert and SRI's first computer mouse prototype.
Above right top Corded computer mouse.
Above right Cordless computer mouse.

Great Inventions We Take For Granted

the paperclip

No.42 the paperclip

"...The paper clip is surely unique, in that people are continually finding so many uses for it other than the one it was invented for..."

In the grounds of the BI Norwegian Business School in Oslo stands a giant 23ft paper clip sculpture. It was erected in honour of an Oslo resident, Johan Vaaler, who for decades was touted as the inventor of the world's first ever paper clip. Whilst he managed to patent his 1899 invention in June 1901 in both Germany and the USA, it is not his design that is celebrated in the sculpture. Vaaler's clips were rectangular and triangular and lacked the crucial last turn and piece of wire that has made double oval-ended paper clip such a useful and well-used piece of stationery.

Nor was Vaaler's paper clip design the first. A flurry of different shapes and designs of bent and shaped wire were developed at the end of the 19th Century and the start of the 20th. The first to receive a patent is believed to be Samuel B. Fay from Venango in Pennsylvania in 1867. His clip was designed less to hold paper together and more as a ticket fastener to attach a laundry or tailoring ticket to especially delicate fabrics and clothing. As Fay cited in his patent, "My invention is especially adapted to securing tickets to silks, lace, and all the finer class of goods, without injury or leaving the slightest trace upon the goods when removed."

Paperwork was tied with ribbon or twine, bundled into folders or sometimes spiked on a wire pin before the invention of these clips, but some designs worked better than others with a number of designs leaving two exposed wire ends that could gouge and tear into documents. Despite lots of digging by technology historians, no one knows who first invented the iconic 'Gem' design with its double-oval shape and rounded ends.

We do know that American William Middlebrook from Connecticut patented a machine for manufacturing Gem-styled clips in 1899 but the design is thought to precede his patent. Apart from their role in stationery, paper clips are used to unclog salt shakers, as emergency hair clips and bookmarks and to access the eject button on CD-ROM drives or to reach the SIM card in certain smartphones. Whilst multiple paperclip designs exist today, manufactured in plastic and plastic coated forms as well as pure metal, as Peter Brown writing in Scientific American in 2009 stated, "to most people, the Gem simply is the paper clip. It's as frozen into office culture as the "qwerty" keyboard."

Above from left to right: A simple solution for a missing cufflink. Formed into a chain link, this paper clip offers a quick fix for broken jewellery. Some simple locks can be unpicked using a metal clip. A bent paper clip makes an impromptu smartphone stand propping the phone up on a table for viewing a video.

Great Inventions We Take For Granted

The supposed inventor of the sandwich was John Montagu, the 4th Earl of Sandwich. It typically consists of cold meat, or cheese and any number of vegetables, relish, pickle, mustard or mayonnaise, conventionally laid between two slices of bread. Usually buttered. It began life in fashionable London society around the gaming tables of the 18th Century, though over time it has become prevalent worldwide. The form supports infinite variations.

the sandwich

The sandwich is named after its supposed inventor, John Montagu, the 4th Earl of Sandwich. It typically consists of cold meat, or cheese and any number of vegetables, relish, pickle, mustard or mayonnaise, conventionally laid between two or more slices of bread, usually buttered. It is thought to have begun life in fashionable London society around the gaming tables of the 18th Century, though over time it has become prevalent worldwide. The form supports infinite variations.

John Montagu inherited his Earldom from his grandfather at the tender age of ten. Throughout his life he would hold a number of important posts in British government and administration including Postmaster General in 1768, Secretary of State in 1770 and the head of Britain's naval fleets as First Lord of the Admiralty, a post he was appointed to on three separate occasions. He proved a great supporter of Captain James Cook's pioneering voyages to the Pacific and in return Cook named the Sandwich Islands after him. He was also something of a player, indulging in early forms of cricket as a young man and extremely fond of long bouts of gambling and gaming, particularly cribbage.

According to frequently recounted stories, around 1762, the Earl was in the middle of a 24 hour card playing binge when he ordered food that could be eaten with one hand, without using a fork and without getting his cards greasy from meat stains. He received a piece of salt beef held between two pieces of toasted bread. Others are said to have started to ask for food, "just like Sandwich's" and the name stuck.

Whilst some elements may be true, and etiquette and elaborate table settings of the time could certainly mean a considerable delay if ordering a full meal, the Earl may have been engaged in important business or work and the story of his continual gambling was potentially nothing more than gossip put about by rival politicians. There is also a very strong case for locating the original home of the sandwich further south and east of England to the eastern Mediterranean region where for many centuries flatbreads had been folded and used to pick up meats, dips and other foods in between the folds.

What is certain is that the story, the name and the food itself became cemented in the British consciousness, and became a popular food in the early years of passenger railways in the UK where sandwiches

No.43 the sandwich

***"The Wall Street Journal* reckons the sandwich may actually be Britain's best contribution to world gastronomy..."**

proved a practical and portable food for passengers. The British love affair with the sandwich continues unabated with sales of pre-packaged sandwiches alone worth in excess of $8 billion per year. The United States adopted the sandwich as their own in the 19th Century, with cookbook recipes and suggestions appearing early in the century and with later inventions such as pastrami on rye, the Reuben and the New Jersey sloppy joe gaining iconic status. Billions of sandwiches are sold every year in the US. According to the What We Eat In America survey, on any given day, 49% of Americans over the age of 20 eat at least one sandwich – hot or cold, toasted, grilled or on fresh bread.

Toasting Auction Success / The World's Most Expensive Sandwich

Part of a sandwich's usual appeal is its affordability and freshness. Neither of these attributes were evident in a grilled cheese sandwich that Diane Duyser, from Florida, stored for ten years, starting in 1994. The grilled cheese sandwich featured a toasted pattern on its outside reminiscent of the Virgin Mary and was sold on eBay in 2004 for a tasty $28,000.

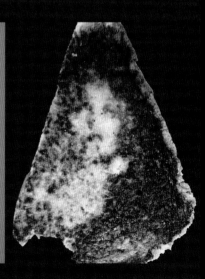

Left page Sandwiches have the capacity to hold a seemingly endless variation of ingredients as fillings.
This page (top left) Traditional English afternoon tea with small triangular sandwiches.
This page (top right) Portrait of John Montagu, 4th Earl of Sandwich by Thomas Gainsborough.
This page (above) The grilled cheese sandwich which sold for $28,000 on eBay.

Great Inventions We Take For Granted

> "I remember I was thinking about dots and dashes when I poked my four fingers into the sand and, for whatever reason—I didn't know—I pulled my hand toward me and I had four lines. I said 'Golly! Now I have four lines and they could be wide lines and narrow lines, instead of dots and dashes. Now I have a better chance of finding the doggone thing."

Norman Joseph Woodland, recounting his moment of inspiration in Smithsonian Magazine, 1999

the barcode

Ubiquitous on store products, those simple series of black lines enable checkout staff to scan and price objects automatically and stores to update their inventory, record transactions and trigger reordering all in real time automatically.

It all started with an overheard conversation. Graduate Student, Bernard Silver listened to a food-chain supermarket executive talking to a dean at Drextel Technology Institute in Philadelphia bemoan the lack of data tracking of goods sold at his stores and how he was searching for a way to speed up the checkout process. Silver contacted a friend and former graduate of the institute, Norman Joseph Woodland, and the pair began brainstorming.

It was whilst reclining on Miami Beach in 1948 that Woodland's mind wandered to the system of dots and dashes that comprised Morse code which he had learned as a boy scout. Idly drawing lines on the sand, he realised that a series of narrow and wide spaced bars could easily convey a code unique to a particular product. The pair worked on devising both a code system and the means to read it, fashioning a ramshackle prototype using a 500 watt incandescent bulb and reading the output via an oscilloscope.

Woodland and Silver received a patent for their prototype bar coding system in 1952 (US Patent 2,612,994) but struggled to find any major corporation that would take their system on. A similar process, KarTrak, was trialled to scan multicolor reflective codes to keep track of rail cars in the 1960s but was abandoned before in 1973, many major grocery and technology companies came together to form the Uniform Product Code Council. This body agreed on a bar code standard – a Universal Product Code (UPC) comprising a rectangular bar code panel and unique 12 digit number as well as implementing the technology to deploy it.

A quite different Miami would finally be the scene for the bar code's commercial debut. The small town of Troy in Miami County hosted a Marsh Supermarket where cashier Sharon Buchanan became the first store clerk to checkout a product using a barcode and scanner developed by NCR on June 26th 1974. The first customer was the supermarket's head of research and development and the first product scanned was a multi-pack of Wrigley's Juicy Fruit chewing gum (which ended up as an exhibit at the Smithsonian Museum). It had been chosen to demonstrate just how small bar

100

No.44 the barcode

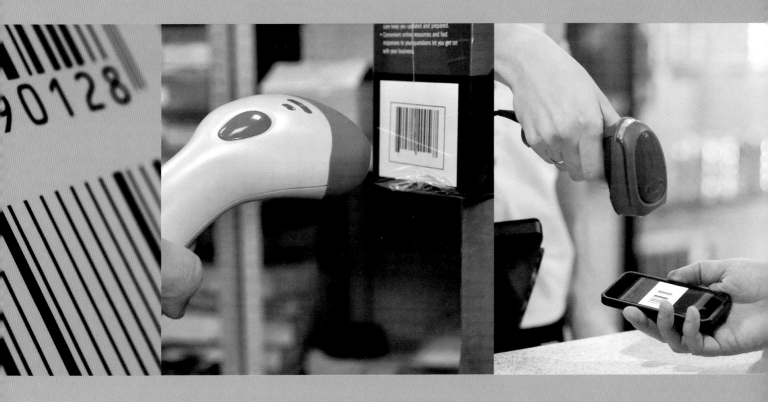

Above (center) A product's barcode is scanned to give details of price. **Above (right)** A retail employer scans a bar code displayed on a customer's phone. **Right** A QR (Quick Response) code, first developed in the 1990s for use in the Japanese automotive industry.

codes could be printed yet still work in the fledgling scanning system.

Other systems such as QR codes have been introduced but UPC barcodes are omnipresent in stores in the US, Canada and elsewhere. They are sometimes scanned by running the product over a glass window of an embedded checkout scanner or the checkout assistant uses a handheld scanner or wand. Apps on smartphones can use the device's camera to scan barcodes and perform online searches for reviews or price matches. Away from stores and products, barcodes are used in a number of other spheres, for example, to record medical samples and in logistics to track packages through their various stages of transport.

101

Great Inventions We Take For Granted

"Throughout history, most inventions were inspired by the natural world. The idea for the pitchfork and table fork came from forked sticks; the airplane from gliding birds. But the wheel is one hundred percent homo sapien innovation."

Megan Gambino, Smithsonian.com, 2009

the wheel

Few inventions have proven more fundamental to advancing humankind's quest to build, explore, and improve, than the humble wheel.

This boon to travelling faster and further over land also forms the basis of gears, water wheels, propellers and turbines.

The first wheels were not used for transport but to work clay. Potters wheels made of a slab of wood or stone lain horizontally and spun by hand were used in Mesopotamia (present day Iraq) around 3,500BCE to turn pots and bowls more quickly and effectively than building up a pot via individual strips of clay.

Historians believe that several centuries elapsed before people had the inspiration to turn a circular slice of wood upwards on its side and use it to reduce greatly the sliding friction encountered in dragging things along. The inspiration may have come from log rollers which saw widespread use in conveying stone blocks for construction, but the ingenuity came in developing a practical axle which enabled wheels to spin freely about their axis to enable easier transportation.

The first wheeled vehicles may have originated in Mesopotamia or the Eurasian Steppes; discovered in Slovenia in 2002, the Ljubljana Marshes Wheel is the oldest existent wheel in existence, measuring 28 inches in diameter and dated to 3150BCE. Early wheeled transportation were simple wooden platforms acting as carts or wagons, pulled by oxen or other draft animals and rolling on a pair of solid wood wheels. Around 4,000 years ago, the first wheels fashioned from a hub and rim connected by spokes were developed. Strong but lighter in weight, spoked wheels offer faster transportation and were first used in war chariots pulled by one or more horses for the first time around 3,700 years ago by North African and Middle Eastern cultures like the Hyskos and the Hittites.

Left side A wooden spoked cart wheel - a design which barely changed for many centuries. **Above (left)** An old potter's wheel. **Above (center)** Modern day car wheel shod with an inflatable tyre. **Above (right)** Primitive wheel on a horse drawn cart in Mongolia.

Great Inventions We Take For Granted

A lunchtime companion for millions of Americans, the good old vacuum flask keeps that coffee hot – or that refreshing ice tea cool, for hours.

the thermos flask

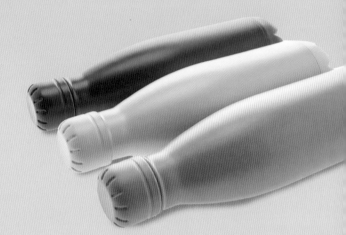

No.46 the thermos flask

"Thermos unites summer and winter, as it keeps any liquid or solid hot without fire and cold without ice until wanted for use."
print advert from the American Thermos Bottle Company, 1921.

These insulating devices essentially consist of two flasks, one inside the other, but joined at the neck with the gap between the two flask walls evacuated of most of the air that would normally lie between. The partial vacuum created prevents heat transfer by either convection of conduction, meaning heat loss or gain for the contents held in the inner flask is low and slow.

The flask was not invented as a lunchtime aid but as a cryogenics tool by Scottish chemist and physicist, James Dewar. Something of a scientific polymath, Dewar's experiments and research extended into astronomy, atomic physics, spectroscopy and invention. In 1889 he worked with Frederick Abel to invent the explosive, cordite – a smokeless alternative to gunpowder. He is best known for his work on liquefying gases, being the first to produce liquid oxygen in large quantities and the first, in 1898 to liquefy hydrogen.

Cooling liquids and gases to extremely cold temperatures was an extremely expensive set of tasks at the time, so six years earlier, Dewar developed a container to store very cold liquids with as little warming through heat transfer as possible. He initially trialled boxes insulated with cork and crumpled newspapers before designing a double-walled flask made of glass with air evacuated between the inner and outer flasks. Dewar's flask which measured 14 inches long and just over five inches wide was exhibited at the Royal Institution in London for the first time on Christmas Day, 1892.

Dewar didn't successfully secure a patent for his flask invention and a partnership of German glass blowers, Burger and Aschenbrenner, became the first to manufacture it for commercial sale, beginning in 1904. They named their product Thermos after the Greek word for heat. An American visiting their Munich-based manufacturing facility, William B. Walker, was so enamoured by what he saw that he bought the licence to produce the flasks in the United States, founding the American Thermos Bottle Company in 1907 and initially producing the flasks from a small factory in Brooklyn.

As sales boomed, Walker's company moved to new premises elsewhere in New Yotk and then in 1913, out of town. A group of 100 prominent citizens from Norwich, Connecticut, calling themselves the Norwich Boomers sold badges at one dollar a time and held fundraising balls and events to produce the capital to entice Thermos manufacture to their town. The resulting business, on the banks of the Thames river in Norwich, produced vacuum flask bottles nationwide between 1913 and 1984. Every crew member of USAF Boeing B17 Flying Fortresses, for instance, were issued a thermos for long bombing missions during World War II. After the war, the boom in school lunchboxes and kits saw the company produce their first celebrity-endorsed model in 1953 with TV cowboy Roy Rogers illustrated on the flask's sides. It was an instant success with more than two million sold in the first year alone.

Stunt Marketing
William Walker proved a shrewd marketer of his product, making extravagant claims for its abilities in the press and displaying a thermos bottle-shaped cars at automotive parades and events. He also ensured newsmakers from President William Howard Taft to polar explorers such as Robert Peary and Ernest Shackleton were all equipped with his vacuum flasks for their newsworthy expeditions.

Left page (above) Three colored vacuum flask bottles. **Left page** A mug, flask and jug thermos. **Above** A vacuum flask dispenses hot coffee at a picnic. **Above (inset)** Traditional thermos flask with plastic finish and screw on cup.

Great Inventions We Take For Granted

the wheelbarrow

The humble wheelbarrow is part of every small construction site and most medium or large gardens. These invaluable one-wheeled transporters work as a second order lever enabling a single user to lift its arms and transport heavy loads from cement or bricks to soil or a heavy crop of vegetables with ease.

The wheelbarrow's origins are shrouded in history with some claiming, one wheeled goods carriers called monokyklos originated in Greece around 2,400 years ago. However, we have to head east to China for the first definitive proof of wheelbarrows' existence by the start of the 2nd Century CE. Some sources grant the title of inventor to the ruler of the Shu Han kingdom of ancient China, Zhuge Liang. According to accounts he introduced the device to his armies as a way of transporting supplies rapidly between battles.

Like many ancient Chinese inventions, the wheelbarrow took its time to trundle westwards and reach Europe. The earliest representations of wheelbarrows in Europe are found in manuscript illustrations in the early 13th Century. Around the same time, a stained glass window fitted in Chartres Cathedral depicts a wheelbarrow whilst across the English Channel, a document from 1222, pertaining to the court of King Henry III states how eight wheelbarrows were ordered for the port of Dover, according to Andrea L. Matthies in The Medieval Wheelbarrow.

The European wheelbarrow placed the wheel at the very front, allowing it to be tipped up and down very easily and to pivot on its single wheel in straightforward fashion. Back in China, more and more wheelbarrows placed an increasingly large wheel underneath the centre of the machine. Whilst making them less manoeuvrable in a tight space, this alteration did make the device easier to propel over long distances, especially when a cloth sail was fitted to the top of the barrow to catch the wind. Sailors and traders from Europe who visited China in the 18th and early 19th centuries were astonished at the speed at which such conveyances could travel.

Sails didn't catch on elsewhere and other than a switch from wooden frame to steel or other metals and increased use of plastics for some of its parts, wheelbarrows have barely altered in design to the present day. The one major exception was the development of a new model by cyclonic vacuum cleaner inventor, James Dyson, in the 1970s. His Ballbarrow replaced the metal or wooden body with a deep well made of injection moulded plastic and took away the front wheel in favour of a large plastic ball. The ball's greater surface area in contact with the ground, helped spread the load and helped made the wheelbarrow more stable over uneven ground.

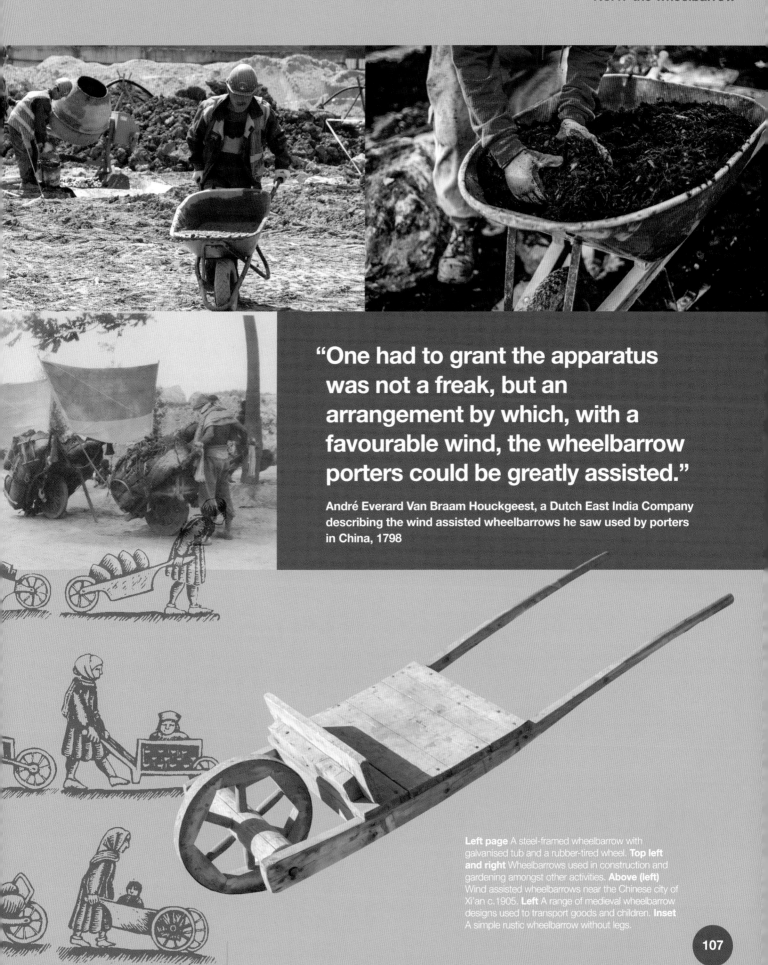

No.47 the wheelbarrow

"One had to grant the apparatus was not a freak, but an arrangement by which, with a favourable wind, the wheelbarrow porters could be greatly assisted."

André Everard Van Braam Houckgeest, a Dutch East India Company describing the wind assisted wheelbarrows he saw used by porters in China, 1798

Left page A steel-framed wheelbarrow with galvanised tub and a rubber-tired wheel. **Top left and right** Wheelbarrows used in construction and gardening amongst other activities. **Above (left)** Wind assisted wheelbarrows near the Chinese city of Xi'an c.1905. **Left** A range of medieval wheelbarrow designs used to transport goods and children. **Inset** A simple rustic wheelbarrow without legs.

There are approximately 900,000 elevators in America today. These travel a total of 1.36 billion miles per year – farther than Earth to Saturn – according to the National Elevator Industry Inc.

the elevator

From the moment the ancient Greeks and other civilizations discovered the pulling power of a rope wound round a compound pulley, people have been able to raise large weights and loads by using pulleys to magnify force. The first elevators were powered by servants or animals pulling ropes which hoisted a platform up a building or structure. French king, Louis XV had a human-powered personal elevator installed in his apartment in Versailles in 1743 to connect his rooms with those one floor above where his mistress, Madame de Chateauroux, resided.

The 19th Century saw the first attempts to fully power and mechanise such lifts or vertical railways as they were often referred to. Steam power was employed to raise and lower platforms in mines and factories with Thomas Hornor and Decimus Burton's 'Ascending Room' amongst the first, in 1829, to offer trips for passengers. They paid to obtain a thrilling view of the London skyline from a platform 121 feet above the ground.

Other systems were developed including an innovative vertical screw design by Otis Tufts. Fitted inside the seven storey Fifth Avenue Hotel in New York in 1859, this complex machine reflected Tuft's fear of hoists and a passenger compartment dangling from ropes or as he put it an, "unconquerable dread and distrust of the principle of suspension." Tuft's machine involved a giant screw extending the entire height of the elevator shaft around which the carriage or car rotated like a nut on a bolt thread. Whilst it worked with few problems for 15 years, it proved impractical to build such designs with much greater vertical travel in order to service the crop of increasingly tall buildings springing up in American cities.

Vermont-born Elisha Grave Otis was working in New York City as a machine installer in a bedstead making firm, Maize & Burns. Whilst installing a machinery hoist at the company around 1852, Otis devised and built a safety system involving a strong wagon spring through which the hoisting cables ran and a toothed ratchet which would mesh and grip the elevator compartment in the advent of the elevator's suspension ropes breaking or giving way.

Otis formed a company with his sons and courted publicity with a daring stunt at the 1853-54 Crystal Palace Exposition in New York. In an elevator shaft open sided so that a large audience could watch, Otis ascended the shaft on a hoist platform before ordering the hoisting cable to be cut with an axe. The platform fell just a short distance before his safety system arrested the fall.

No.48 the elevator

The demonstration achieved its target, convincing enough industrialists that elevators could be safe enough to carry customers and residents. The first commercial passenger elevator inside a building was installed in 1857. The high-end Haughwout Department Store in New York City, on the corner of Broadway and Broome, featured an Otis-designed elevator powered by a steam engine in the store's basement. It ascended the building at a slow rate of eight inches per second but proved a novelty and attracted customers into the store. Otis received a patent for his Improved Hoisting Apparatus in 1861 and twelve years later, there were some 2,000 Otis elevators in use. The company expanded overseas and won the commissions for elevators in the Eiffel Tower in Paris, France as well as, back at home, in both the Flatiron Building and the Empire State Building.

Left page Inside the shaft of a glass elevator.
Above Elisha Otis demonstrates his freefall prevention system in New York City in 1854. **Top right (above and below)** An antique elevator floor indicator. An antique wooden elevator inside a metal safety cage.

Shanghai Tower in Pudong district, Shanghai.

The fastest passenger elevator currently installed is found in the 2,073 feet-tall Shanghai Tower in the Pudong district of the Chinese city. It reaches a top speed of 67 feet per second or 46 miles per hour. At this rate, it can travel from ground level to the 119th floor in just 55 seconds.

Great Inventions We Take For Granted

cat's eyes

Innumerable lives have been saved through a simple yet ingenious device invented in the 1930s by a former boiler maker from Halifax, England, who was working for himself, repairing roads and garden paths using tarmac, when he had his spark of inspiration.

Percy Shaw had always been resourceful, inventing games as a child and a new form of petrol pump as a young man. Trams used to travel down the same roads as Britain's relatively few cars in the 1910s and 1920s and drivers had got used to guiding themselves at night by the tram lights in lanes to the side of their vehicle. The roads near to where Shaw lived had had their tram lines removed and with little street lighting, staying on the correct side of the road whilst driving at night proved hazardous, especially in foggy or rainy conditions.

One night in 1933, Shaw was driving home from the neighbouring village of Queensbury, descending down a twisting road. A sharp reflection in his headlights stirred his curiosity and caused him to brake and get out of his car. He discovered that this reflection was the eyes of a cat and that he had been driving down the wrong side of the road. If he had continued his trajectory, he would have crashed off the road and plummeted down the hill. Shaw had also noted the reflective markers fitted to some road signs at eye level which bounced motor vehicle headlight beams back to illuminate the sign. He reasoned that these reflectors were needed down the middle of roads, especially those lacking street lighting.

After a year of trial and error, Shaw took out a pair of patents for his reflecting road studs (colloquially known as cat's eyes or catseyes) in April 1934 and in March 1935 Reflecting Roadstuds Ltd was incorporated, with Shaw as Managing Director and just 500 UK pounds ($2,500 at the time) of financing. Major impetus for the product came during World War II and the blackout imposed in much of Britain to reduce light that could be a navigational aid to enemy aircraft. The company expanded and began producing more than a million cat's eyes a year.

Shaw's device was a study in simplicity and ingenuity. Four bullet shaped glass beads, two facing in each direction, were housed in a metal base and surrounded by a robust and resilient flexible rubber dome. When fitted every few paces in series down the middle of a road, they reflected vehicle headlights to clearly demarcate the two sides of the road without the need for a power source. When a passing vehicle drove over the device, the dome flexed and pushed the glass beads downwards to stop them breaking. In later, self-cleaning versions, the base would allow rainwater to collect so that when depressed, the beads were washed and wiped by the rubber housing as it flexed and unflexed.

No.49 cat's eyes

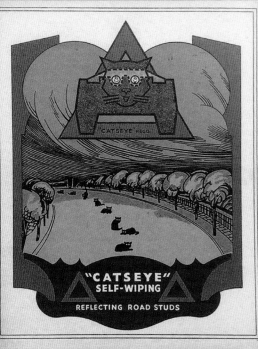

Left page Botts dots lane markers are embedded into an American highway. **This page (above)** Cat's eye road studs in their housing lying above the road surface. **Above** A late 1940s promotional poster and catalogue image for Shaw's cat's eyes. **Right** The twin reflective spheres shown set into a cat's eye **Inset** Percy Shaw.

Shaw became wealthy from his invention, but remained living in the same modest house, from which much of the furniture and curtains were removed, yet four televisions were constantly turned on and tuned in. Parked outside his home was his one sole extravagance – a Rolls Royce Phantom luxury car.

Great Inventions We Take For Granted

"Today, if all of them in a single neighborhood were started at once, the racket would be heard round the world."

Lawns expert C.B. Mills about gasoline-powered motor lawnmowers in the United States in 1961.

the lawn mower

Handheld scythes were the most common method of cutting grass, other than letting grazing animals such as goats and sheep loose on your lawn, before an English engineer, Edwin Beard Budding, hit upon a mechanical rotary cutter in 1830. Budding from Gloucestershire in England had viewed a machine at a textiles mill using a cutting cylinder to trim cloth after it had been woven and figured that a similar device pushed over ground could be used to trim grass. Venturing into partnership with John Ferrabee, the pair built cast iron lawnmowers, around 19 inches in width with a large roller at the rear and a cutting cylinder or reel at the front. As the mower was pushed forward and the roller turned, cast iron gears, transferred the movement to turn the cutting cylinder to trim the grass.

Other companies developed Budding's design, adding sidewheels which proved popular across the Atlantic where grass was often coarser. The push lawnmower received its first major competitor in the 1890s with the arrival of motorized lawnmowers. Some of the first were powered by steam but were quickly superseded by petrol-fuelled internal combustion engines. Using a patent by automotive pioneer, Ransom Olds, the Ideal Power Lawn Mower company started selling motorized lawnmowers in 1914 and eight years later sold the first ride-on law tractor in the US, the Triplex.

In 1963, and inspired by the British engineer, Christopher Cockerell's invention of the hovercraft, Karl Dahlman invented the first hover mower. A fan forced air downwards to raise the lawnmower's lightweight body, letting it float on a cushion of air whilst a blade span at high speed to cut the grass. Dahlman's Flymo machines first rolled off production lines in England two years later and the company was acquired by Electrolux in 1969.

No.50 the lawn mower

"A new combination and application of machinery for the purpose of cropping or shearing the vegetable surfaces of lawns, grass-plats and pleasure grounds. Country gentlemen may find, in using my machine themselves, an amusing, useful and healthy exercise."

extract from Budding's patent, October 1830.

According to the Consumer Product Safety Commission more than 80,000 Americans visit the Emergency Room each year due to lawnmower-related injuries. The Amputee Coalition cites that 20,000 people are injured by 'ride-on lawnmower' accidents, including 800 children, three quarters of which lose a limb as a result.

Left page Robomow City 110 robotic lawn mower cuts lawns with no human supervision. **Left page (inset)** An early reel lawnmower was pushed by hand with the cutting blades turned by the rotation of the mower's wheels. **Above left** A petrol powered lawnmower features a safety bar which must be gripped for the motor to run. **Above (top left)** A ride on mower for use over larger areas. **Above (top right)** Modern mowers allow adjustable cutting height to tackle shorter or longer grass. **Left** A gasoline-powered walk-behind mower with fitted cuttings hopper.

Great Inventions We Take For Granted

the magnetic compass

North, South, East or West, if you need to know your way (and have left your smartphone or GPS receiver at home) a compass is still best.

Early explorers and sailors navigated by sight, seeking out recognisable landmarks, following flight patterns of birds or using the position of the Sun by day and the stars in the night sky to find their way. It proved a risky business and much shipping rarely ventured too far from a viewable coastline for security.

Some point around 200BCE, it is thought that the Ancient Chinese discovered how magnetite, a common iron ore, possessed magnetic properties and that these could be used to orientate one's position. Magnetite (also known as lodestone) is the most magnetic naturally-occurring mineral on Earth. Its property of attracting iron objects was noted by the ancient Greek philosopher, Thales of Miletus around 580BCE, but it was in China that magnetite was first suspended or floated, so that it aligned with Earth's magnetic poles and was used for divination and fortune-telling as well as orientating builders so that their constructions faced a certain direction to assure good fortune.

At least eight centuries passed before the first known uses of magnetite as a form of direct, travelling navigation aid. In 1088, Song Dynasty scholar Shen Kuo wrote that when, "magicians rub the point of a needle with lodestone, then it is able to point to the south". Needles, nails and other small pieces of iron were magnetised in this way and then floated on straw or another buoyant substance in water or hung from a strand of silk attached by wax, to form the first navigational compasses. Such devices reached Europe several centuries later, possibly through interactions between Asian, Arabic and European traders and sailors. At some point in the 13th Century, the first dry compasses were adopted in Europe, with a magnetised needle on a pivot so that it could turn freely to point northwards, and attached to a compass card displaying the differing directions all housed in a wooden box. Adopted by pioneering explorers like Christopher Columbus and Vasco Da Gama to discover new lands, these compasses were the forerunner of today's handheld devices used by hikers, sailors and those engaged in orienteering.

No.51. the magnetic compass

"Magnus magnes ipse est globus terrestris - The whole Earth itself is a magnet"

William Gilbert, English physician in his book, De Magnete, 1600CE.

A range of pocket compasses, old and new, and fashioned from metal or plastic all provide invaluable navigation assistance.

Great Inventions We Take For Granted

The humble umbrella's name derives from umbra, the Latin for shade or shadow, hinting at how these rainstoppers began life as sunshades.

the umbrella

Umbrellas developed out of parasols and sunshades used as far back as the time of the pharaohs 3,500 years ago. Ancient Egyptian sunshades made of wood, animal bone frames covered in palm leaves and other plant matter were often wielded by servants and slaves to keep nobles shaded and cool. Other ancient civilizations including the Greeks, Chinese and Indians employed similar devices as sunshades, the Chinese replacing palm leaves with canopies made of silk or paper. China was also the home of the first collapsible parasol made of oiled paper and bamboo. The handle and staff of the device was hollowed out allowing it to collapse and open in a similar way to a spyglass or small telescope.

A revival in parasol use by female members of the gentry in Europe began in the late 16th and 17th centuries. These models, often with whalebone bibs and canopies woven from silk, provided minimal water resistance although when the silk was oiled it offered some protection against a downpour. New materials such as waxed cotton and, later, nylon and other synthetic fibres, have since replaced silk to provide genuine water protection.

The umbrella's status as solely a female fashion accessory gradually changed as they were popularised by figures such as French scientist Navarre, who demonstrated a push button opening design in the 1780s and the founder of the Marine Society and Magdalen Hospital in London, England, Jonas Hanway who from the 1750s onwards always carried an umbrella around London regardless of any ridicule he suffered. New designs continued to appear, some replacing natural materials with steel for the ribs and shaft with German, Hans Haupt's invention of the pocket umbrella in 1928 especially influential.

The umbrella is one of the most common objects for new patent submissions, so much so that for a time the US Patent Office employed dedicated staff to sift through the applications. As author Susan Orlean wrote in The New Yorker in 2008, umbrellas are, "so ordinary that everyone thinks about them, and, because they're relatively simple, you don't need an advanced degree to imagine a way to redesign them, but it's difficult to come up with an umbrella idea that hasn't already been done."

Above Detail of a paper umbrella.
Right page Commuters using their umbrellas on a rainy day.
Right Parisians in the rain with umbrellas, by Louis-Léopold Boilly (1803).

No.52 the umbrella

"Those who do not want to be mistaken for vulgar people much prefer to take the risk of being soaked, rather than to be regarded as someone who goes on foot; an umbrella is a sure sign of someone who doesn't have his own carriage."

a Paris magazine in 1786 remains unimpressed by the use of Umbrellas.

In the 1920s, Baltimore was the centre of global umbrella making, producing over two million a year. The Chinese city of Songxia now holds the title with around 1000 factories which produce upwards of 500 million umbrellas annually.

Great Inventions We Take For Granted

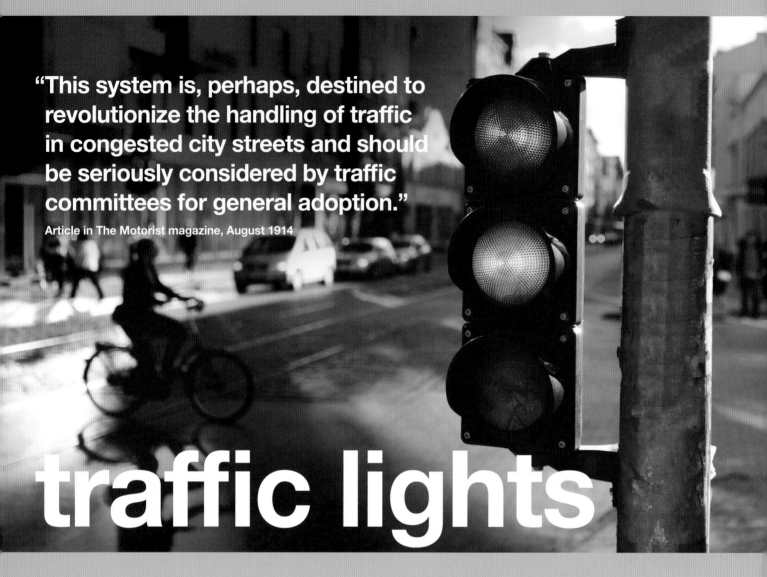

"This system is, perhaps, destined to revolutionize the handling of traffic in congested city streets and should be seriously considered by traffic committees for general adoption."
Article in The Motorist magazine, August 1914

traffic lights

A first come, first served policy at junctions and intersections proved chaotic and dangerous for the booming number of horse drawn vehicles, bicycles and velocipedes and later, motor vehicles, in rapidly growing cities. Traffic lights are often unheralded, uncelebrated and more often roundly cursed when progress is halted by a red light or a ticket received for running one, but they have proven significant lifesavers.

The world's first set of traffic lights were installed outside Britain's Houses of Parliament in London on the 9th of December, 1868. They lasted less than a month. The 22 feet tall post located on the north-east corner of Parliament Square (a spot now occupied by a statue of British prime minister, Winston Churchill) featured two gas powered lights, a red one for stop and green for go, as well as two semaphore arms that were cocked or straightened to tell road users how to proceed. The design by John Peake Knight, was heavily influenced by the railway signalling of the time yet still needed a policeman in attendance to operate the signal. On the 2nd of January the following year, an explosive accident with the gas supply to the lights injured the attendant officer. Traffic lights didn't return to London until 1925.

In the United States, semaphore traffic signals were widely in use in America's increasingly populous cities, yet despite the motor vehicle being in its infancy, more than 4,000 people died in car crashes in the United States in 1913, according to the Smithsonian which stated, "when those unforgiving machines met at a crowded intersection, there was confusion and, often, collision." Police officers, charged with directing the traffic were often at risk and frequently ignored. One Utah law enforcer, Lester Wire, who worked as a detective on the Salt Lake City force, is credited with developing the first electric traffic light around 1912. His design was mounted on a 10 foot pole and featured light bulbs painted green and red on all four sides of its wooden box with a heavily sloped roof. Salt Lake City would become the first city to run interconnected traffic signal systems when six intersections received traffic lights all controlled from a central manual switch in 1917.

Seventeen hundred miles east, Cleveland engineer, James Hoge took railway signals as his inspiration but cleverly tapped into the electricity cabling that served trolley car lines running along city streets. His system debuted in Cleveland at the intersection

No.53 traffic lights

between 105th Street and Euclid Avenue in 1914, a year after he made his patent application and four years before it was granted. His signal lights illuminated signs displaying the words, "stop" and "move" and this "municipal traffic-control system" was the first that could be paused when emergency vehicles needed to get through. Hoge's system and variations thereof were widely adopted whilst automatic traffic lights that dispensed with the need for a human operator at each junction were first patented in 1923 by another Cleveland resident, Garrett Morgan, one of many patents the inventor of the smoke hood and other devices obtained during his lifetime.

Left page Lights attached to a pole turn amber beside a highway. **This page (left)** Advert for an "Electric Traffic Regulator" in the Pryke & Palmer catalogue of 1930. **Top (right)** Three sets of traffic lights inform drivers at a busy junction. **Center (right)** Hoods over the tops over each light prevent the Sun's reflection giving off a phantom signal. **Below (right)** Semaphore lamp signal on the railway cuts through the gloom. **Below** A vintage illuminated American traffic light.

Detroit police officer William Potts was the first to propose and build a traffic signal in 1917, incorporating an orange or amber light to ease the transition between red and green lights.

Great Inventions We Take For Granted

Every time you buckle up as a passenger or driver, you are drastically reducing your chances of death or serious injury thanks to this restraining invention.

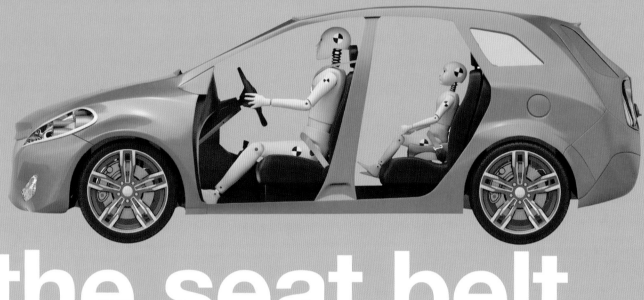

the seat belt

The first restraint to keep a person in place or in their seat was invented by George Cayley to keep pilots of his 19th Century experimental gliders in position. These were quickly followed by Edward J. Claghorn's 1885 invention of a seat belt for passengers taking trips along New York City's frenetic streets in a horse-drawn taxi. Whilst automotive design advanced rapidly in many aspects, driver and passenger safety lagged behind, so much so that both the rear view mirror (in 1911) and the car seat belt (in 1922) actually made their debuts at the famed Indianapolis 500. For the 1922 race, Barney Oldfield, the first man to run a 100mph lap at the Indianapolis Motor Speedway track, was driving the pace car and fitted the vehicle with a modified parachute harness as a restraint.

Lap or two point seat belts were fitted to a handful of vehicles in the 1930s and companies like Nash, Ford and Chrysler offered them as options in the 1950s. These consisted of a strap across the waist or lap and did little to prevent a victim's upper body being flung towards the windscreen or dashboard on impact. Across the Atlantic at Swedish car-makers Volvo, Nils Bohlin leveraged his experience in designing ejector seats at aviation company, Saab, to build a more effective belt restraint system. He combined a lap belt with a diagonal belt across the chest and anchored the straps low near the base of the seat to increase stability. The belt was fastened using a single buckle and was adjustable in length to suit a range of body sizes.

Bohlin's belt was first fitted to Volvo cars in 1959 but instead of keeping the innovation private, Volvo waived their patent rights, enabling many manufacturers to follow suit.

Three point seat belts became a mandatory fitting in all new American vehicles in 1968 but installing them in cars and getting people to actually wear them proved two different things. Increased state legislation, more safety campaigns and greater public awareness led to a rise in belt use on America's highways from 14% in 1983 to 89.7% in 2017 according to the National Highway Traffic Safety Administration (NHTSA). Whilst enhancements have been introduced including more comfortable belt webbing, tensioners to eliminate slack and force limiters to manage the stresses placed on the user when restrained, many Automotive vehicle's three point belts bear more than a passing resemblance to Bohlin's design and, when used, reduce the risk of death in road traffic accidents by an estimated 45 per cent.

No.54 the seat belt

"Seat belt use is the most effective way to save lives and reduce injuries in crashes."

Center for Disease Control and Prevention (CDC).

Left page Crash test dummies modelled on adult and child proportions sit in a vehicle prior to a crash test. **Above (top left)** A car's three point belts bear much similarity to Bohlin's original design. **Above (top right)** Modern motor vehicles are fitted with audible and dashboard warning signals should seat belts not be worn. **Above (Center)** The highest part of the restraint system is fitted into the side of the vehicle's cabin. A seat belt buckle and tongue. **Above (center right)** Turkish stamp (c.1987) promoting road safety. **Above** Crash test dummies test out both the seat belt restraint system and protective airbags in a new vehicle.

Great Inventions We Take For Granted

"Wouldn't it be great to have a tool that performs like a wrench but can grip like a pair of pliers?"

Dan Brown, inventor of the Bionic Wrench.

the adjustable wrench

No.55 the adjustable wrench

In 2002, Dan Brown invented a new type of adjustable wrench with a pair of handles, like pliers, but a mechanism that closed around a nut or bolt similar to an SLR camera's shutter. The tool, named the Bionic Wrench, grips all six sides of a hexagonal nut and first went on sale in 2005.

NASA is developing 3D printing so that tools, including wrenches, can be constructed on the ISS and future spacecraft.

Left page A well-used adjustable wrench. **Above (top left)** A collection of old steel wrenches of varying sizes. **Above (right)** A wrench's furled control enables its jaws to be closed round a set of water pipes with hexagonal screw fittings. **Above (left)** An adjustable wrench produced as one individual object via 3D-printing.

The ultimate one-size-fits-all tool is also one of the most underrated in the toolbox, yet frequently saves the day by fitting round any size nut or bolt head.

Swedish inventor Johann Petter Johannson would obtain over 100 patents worldwide during his lifetime. None would be more lucrative than those he obtained for his spanner wrenches. As the owner of the Enköpings Mekaniska Verkstad workshop, Johannson faced a profusion of different nut and bolt sizes when conducting his work and often had to forge or make a new wrench each time he tackled a new repair or task. Pondering the problem and the time it wasted, he developed what he initially called an 'iron hand' in 1888. This hinged spanner could accept different size fittings and became known as a pipe or plumber's wrench.

It wasn't the first pipe wrench; Daniel Stillson a tool machinist in Cambridge, Massachusetts had developed one such model back in 1869, but three years later, Johannson refined his wrench design which could struggle with turning stuck or rusted bolts. He added a furled wheel which when turned allowed the wrench's jaw width to widen or narrow as one jaw moved along a screw thread whilst the other jaw stayed fixed. The adjustable wrench could be clamped onto the head of any sized nut or bolt head.

Obtaining Swedish patent no.4066 in May 1892, Johannson manufactured his first generation wrench in six sizes and struck an agreement with Swedish company, B.A. Hjorth & Co to distribute his inventions worldwide under the Bahco trademark. It resulted in more than 100 million of his wrenches being manufactured. Many were used in industry or on railway locomotives; others found their way into home or car toolboxes with myriad uses from assembling flat pack furniture to adjusting the ride height of a bicycle.

Great Inventions We Take For Granted

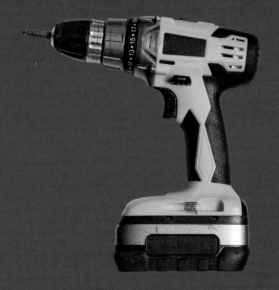

the power drill

The home improvement power tool of choice, and the first item in many new workshops, these versatile DIY appliances put powered woodworking and construction into the hands of ordinary people.

The world's first electric drill was not American, a fact that may shock fans of the heritage of pioneering power tools company, Black & Decker. Scottish-born engineer Arthur James Arnot travelled to Australia in 1886 to build the Sydney Street power station in the city of Melbourne. Working with William Blanch Brain, the pair developed the first drill powered by alternating current electricity in 1889. It was neither portable nor designed for home improvements, more an industrial rock drilling device with uses in construction and mining.

Six years later, German engineers, Wilhelm Emil Fein and his brother, Carl, invented the first portable power drill, although portable was something of a misnomer with the device weighing 16.5 pounds. An operator required both hands to grip the two handles and all their body weight to lean against the drill's chest plate as its relatively weak DC motor span the clamped drill bit.

In 1910, two young friends working at the Rowland Telegraph Company in Baltimore, Maryland decided to set up their own business, opening a small machine shop in the city, deriving start-up funds from one re-mortgaging his home and the other selling his car for $600. S. Duncan Black and Alonzo Decker's first completed drill is now found in Washington D.C.'s National Museum of American History such is its iconic status.

Inspired by a Colt revolver lying around, the pair formulated the pistol shape for the half inch drill with the switch in the trigger position on the handle, enabling easy, one handed operation and control. At the heart of the drill was a universal motor capable of running off both AC and DC power. The drill was launched in 1917, the same year that Black and Decker opened a manufacturing facility in Towson, Maryland. Within two years, the fledgling company reached a turnover of $1 million.

Cordless Drill
Spinoffs from NASA's work with the space program are legion from CMOS image capturing sensors used in digital cameras to scratch-resistant lens and memory foam. Other items, though, from Tang to Teflon, are wrongly attributed as NASA's own work and the same is the case with the cordless drill. In 1961, Black & Decker, now a major

No.56 the power drill

corporation flourishing as part of the home improvement boom it had inspired, introduced the first cordless drill, the C600.

Selling at $60 – around four times the price of a corded model, and with nickel cadmium batteries only able to offer 4.2 volts leading to slower drill speeds, the model was nonetheless successful, particularly with anyone – from boat builders to outdoor repairers – who had to work in or close to water. Black & Decker were contracted by NASA, via Martin Marietta, to produce a cordless drill – the Apollo Lunar Sample Drill or ALSD –which flew on the Apollo 15, 16 and 17 missions and was used to take core samples of the lunar surface.

Top left to right: A cordless drill powered by a battery pack fitted to its handle. A corded drill that plugs into a mains supply. A vintage drill cut away to reveal its windings round its armature. A heavy duty electric drill with hammer drill function and an addition grip near the chuck. **Center left to right** Drilling into a shelf surface. With a crosshead bit fitted firmly in the chuck, this power drill can affix screws quickly and accurately. A high speed drill bit bores a hole in wood gripped in a vice.

> "When I started off, I was working in a shed behind my house. All I had was a drill, an electric drill. That was the only machine I had."
>
> James Dyson, British inventor.

Great Inventions We Take For Granted

> "The pen is an instrument of discovery, rather than just a recording implement."
>
> Billy Collins, former Poet Laureate of the United States, 2001

pens and pencils

Despite electronic media, our reliance on these writing instruments endures, whether it's jotting down a phone number, writing out a shopping list or signing one's name with a flourish.

Pens made of cut reeds or from the hollow stem of a bird's flight feathers and known as quills were the writing instruments of choice for many, many centuries. Thousands of manuscripts were written and important documents like the 1787 Constitution of the United States signed, using quill pens, before a flurry of invention began in the 19th Century. The first innovation to arrive was the dip pen which had capillary channels to the sides of the metal nib allowing a flow of ink to the nib after it has been dunked in an ink well or pot. Fountain pens – a pen with its own internal reservoir of ink – are thought to have been developed in France in the 18th Century and became widely used, especially after Romanian, Petrache Poenaru patented a design with a quill from a swan acting as a small ink reservoir in 1827. Ink flow remained sporadic and unregulated with early fountain pens, resulting in scratchy writing or leaks and blotting until 1884 when Lewis Edson Waterman popularised his own revised version with three channels feeding ink to the nib in a smoother, more continuous fashion.

Ballpoint Pens

Fountain pens still had the power to frustrate those who relied on them for a living, such as Hungarian journalist, László József Bíró. Apart from leaks and frequent refills, he was irked by how the slow-drying ink smudged on the page in comparison to the thicker, oil-based ink used to print newspapers which dried almost instantly. Biro and his brother, György, a chemist, experimented, noting how newspaper ink would not flow from a fountain pen nib well, so developed a pen with a small ball bearing acting as the tip and turning as it moved across the page, thus coating itself with an ink from the pen's internal reservoir.

Such an innovation wasn't totally new; the Harvard-trained lawyer, John J. Loud, for example, received a patent for a similar device in the 1880s but his instrument featured a large steel ball, was developed

No.57 pens and pencils

László József Biró

Biro's ballpoint design was the inspiration for Helen Barnett Diserens to invent the first roll-on ball deodorant. It went on sale as the Ban Roll-On in 1952 in the United States.

A typical ballpoint pen can write 45,000 words before it runs out, according to OfficeXpress.

Above A feather quill pen rests its nib inside an ink bottle. **Right page (top)** left Ink pens used to perform calligraphy. **Right page (top right)** The writing ball of the plastic ballpoint pens are often made from steel or tungsten carbide. **Right** A collection of antique pens and pencils.

Great Inventions We Take For Granted

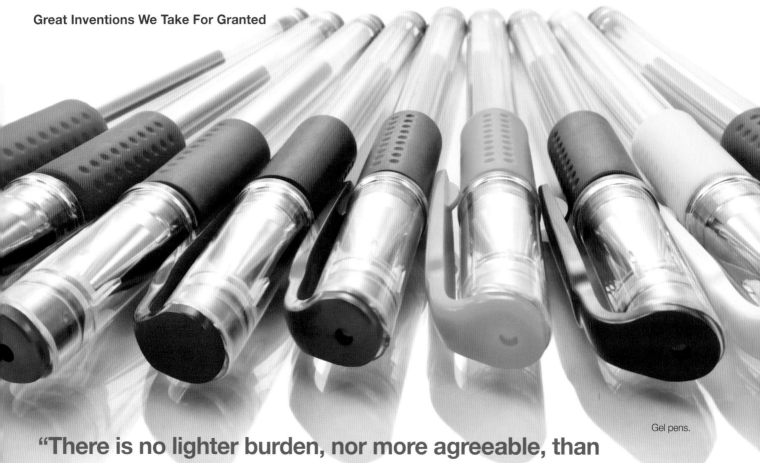

Gel pens.

> "There is no lighter burden, nor more agreeable, than a pen. Other pleasures fail us, or wound us while they charm; but the pen we take up rejoicing and lay down with satisfaction, for it has the power to advantage not only its lord and master, but many others as well."

Francesco Petrarch, 14th Century Italian scholar.

primarily to write on rough surfaces such as leather, and Loud found that it worked poorly on paper and didn't commercially develop the idea further. What was innovative about the Bíró brothers' ballpoint was the combination of a fine tip, small ball bearing and the use of almost smudge-free ink. It proved a winning combination that was patented in France in 1938 and, after the brothers fled Europe during World War II for Argentina, went on sale as the Birome pen. Amongst the first major orders were from the British Royal Air Force to equip their pilots and aircrew. Early adopters reported astonishment at the ease of use and how long the pen lasted without refilling.

In the United States, a ballpoint pen battle raged in stores and courts between Eversharp who paid half a million dollars for the official rights to the Birome, and Reynolds International Pen Co. who went ahead with their own similar design and got it to market first in October 1945. Time magazine reported in October 1945 that, "thousands of people all but trampled one another last week to spend $12.50 each for a new fountain pen. It had a tiny ball bearing instead of a point, was guaranteed to need refilling only once every two years, would write under water (handy for mermaids), on paper, cloth, plastic or blotters." An inexepsnive alternative to both manufacturer's models would eventually be offered by Frenchman, Marcel Bich, who, having licenced the Bírós' technology in 1950 promptly introduced the disposable ballpoint pen. It featured a 0.039 inch diameter stainless steel ball and became known as the Bic Cristal. In 2006, the hundred billionth Bic Cristal was sold, making it the world's best-selling pen.

Gel Pens

Japanese company, Sakura investigated production of new forms of inks at the start of the 1980s, focusing on a unique property of gels called thixotropic action where if a gel is disturbed, its viscosity increases and it becomes more liquid whilst at rest, the gel's viscosity decreases. Seeking the perfect gel ink ingredients, Sakura employees experimented with over 1,000 different ingredients, some as outlandish as grated yam and egg whites before settling on xanthan gum, a food additive found in many instant soups and some jams. The eventual result was a new form of gel ink which was water-based but produced bright colours and was fade resistant. Sakura's first gel pens went on sale in 1984.

No.57 pens and pencils

Lead Pencils

Despite the name, black pencils contain no lead whatsoever but are made predominantly of graphite – an allotrope of the chemical element carbon. When it was first discovered, people thought it was some strange form of the metal, lead and named it plumbago or black lead.

In 1564, a huge deposit of solid graphite was discovered in Borrowdale in north-western England. It puzzled local farmers with its similarity to coal yet lacking that fuel's flammability but it left its mark, literally and was first used by local farmers to mark and identify the sheep in their flocks before later being mined heavily to make moulds for cannonballs. The solid graphite was so valuable that mines were guarded and graphite miners flogged with whips if they were caught stealing any of the precious substance. Small square sticks of graphite were sawn off and then wrapped in string, sheepskin or another covering to make an incredibly useful and versatile writing and drawing tool widespread throughout Europe.

When British ships blocked trade with France in the 1790s so the French could no longer import solid graphite sticks, an enterprising, one-eyed military balloonist and engineer was asked to come up with an alternative. Nicholas Jacques Conté mixed graphite powder with clay, formed narrow rods of the material which he baked in a kiln oven and, following a tradition begun in Germany in the 1660s, placed each rod in two halves of a cylinder made of wood. With these acts, Conté invented the modern pencil. By varying the amount of clay and graphite, Conte discovered he could make pencils of different blackness and hardness. Conte's grading system from 9B (the softest) to 9H (the hardest) is still used in many parts of the world outside the United States.

Pencils were sharpened with a knife until French mathematician Bernard Lassimone patented a pencil sharpening machine in 1828. Across the water, African-American carpenter John Lee Love invented a design of sharpening still in use to this day. The Love Sharpener features an opening to place the pencil nib, a hand crank to rotate the shaving blades and a small compartment to capture the pencil shavings.

Above Newly-sharpened lead pencils.
Below A 'Love'-styled pencil sharpener with a hand crank that turns the sharpening blades.

Great Inventions We Take For Granted

> "They're used all over the Earth, and there are always improvements and additions to the line…It's like having your children grow up and turn out to be happy and successful. When Post-its are still used after I am gone, it will be as if a part of me will live on forever."
>
> Art Frya

the post-it note

From communicating with family members via messages stuck on the kitchen refrigerator to playing an impromptu game of "Who Am I?" with one stuck to your forehead, these sticky notes are ubiquitous in modern life.

The history of invention is littered with accidental discoveries and extraordinary moments where initial disappointment turns to joy and potential dollar signs when an object or substance is re-appropriated and turned to great use for a quite different application. This was the case with bubble wrap, first devised as wallpaper, and with mauveine - the first artificial dye – which a teenaged British chemist William Henry Perkin created whilst trying to concoct a cure for malaria. The same is the case with an adhesive created by San Antonio-born chemist, Dr. Silver Spencer in 1968.

Silver was working at 3M's Central Research Labs when he developed a remarkably low tack adhesive (Acrylate Copolymer Microspheres) made of microscopic acrylic spheres that only stuck lightly to surfaces but didn't bond tight with them. "At that time we wanted to develop bigger, stronger, tougher adhesives," said Silver in a later interview. "This was none of those." The adhesive would stick to paper but could be pulled away again and again multiple times with little degredation in its light grip.

For a number of years, Silver had a solution on his hand looking for a problem. He presented the adhesive at numerous company meetings and proposed its use for bulletin and message boards but his proposals gained little traction. Another 3M scientist, Art Fry, recalled one of Silver's seminars in 1974 when seeking out a way to solve a problem with bookmarks for his church choir. An avid choir singer, Fry used pieces of scrap paper to mark places in his hymnal for the next service only to find his bookmarks fell out. Using Silver's adhesive, Fry was able to produce reusable sticky notes, but it became quickly apparent that these slivers of paper had considerably more potential than just bookmarks. 3M's own offices were used to test out the product before a small soft launch occurred in four cities in 1977 under the name Press 'n Peel. By the time they were sold nationally in 1980, they were renamed Post-it®.

No.58 the post-it note

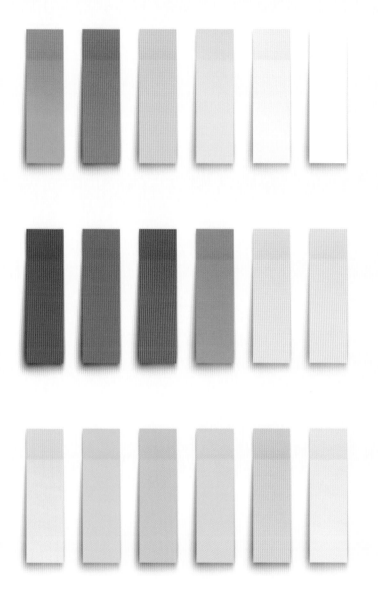

Left page Blocks of Post-It notes. Each note peels away from the note beneath it. **This page** Post-Its produced in a range of sizes and colours can mark places in books, map out a project on a noticeboard or changes in a document or presentation.

The distinctive canary yellow colour of standard Post-it® Notes happened circumstantially as that was the colour of the spare paper in the lab next door to the 3M team working on the product.

Great Inventions We Take For Granted

"Before the Swingline, you practically needed a screwdriver and a hammer to put the staples in. He and his engineers devised a patented unit where you just opened the top of the machine, and you'd plunk the staples in."

Alan Seff, son-in-law of Swingline inventor, Jack Linsky.

the paper stapler

Few office products are more humdrum than the simple stapler, yet, without its satisfying crunch, where would you and your collated documents be?

It is said that the first stapler was fashioned for King Louis XIV in the 18th Century with each staple elaborately produced and bearing his own insignia to act like the royal seal. No trace of the device remains, but the increasing preponderance of paper, offices and administration in the 19th Century prompted the invention of more mundane staplers and other stationary devices like binders, hole punches and paper clips to keep all the pages of documents together in order.

Early staplers just thrust a wire staple through the paper but the wire ends had to be folded over to clinch the paper by hand.

In 1879, George W. McGill introduced his "Patent Single Stroke Staple Press" which clinched a staple through the paper in one simple, thumping action. McGill's device tipped the scales at more than two and a half pounds and only held one half inch wide staple at a time, demanding constant reloading. A Norwalk, Connecticut company, E.H. Hotchkiss Company debuted their No. 1 Paper Fastener around 1901 with a long spiral series of staples fed into the rear of the machine. Hotchkiss staplers became widely used in America and abroad, so much so that in Japan, all staplers today are still referred to as hochikisu.

Swingline, arguably America's most famous staple brand today, started out as the Parrot Speed Fastener Company in 1925, founded by Jack Linsky. In 1939, it introduced a new stapler design which opened at the top, allowing a row of adhered staples to be simply dropped into the device, and held in place by a spring, making life for the office worker just that little bit easier.

Above Short strips of staples.
Right page (main image) The classic design yet relatively newly produced red Swingline stapler. **Right page (far right)** McGill Single-Stroke Staple Press, patented on February 18, 1879. **Right** An invaluable object in any office, a stapler is used to connects pages of a document together.

No.59 the stapler

Beavis and Butthead creator, Mike Judge's 1999 movie, *Office Space* featured a red Swingline stapler that the company didn't manufacture but started producing after the film's cult success. According to Quartz magazine, every US Swingline employee now receives a red stapler when they join the company.

Great Inventions We Take For Granted

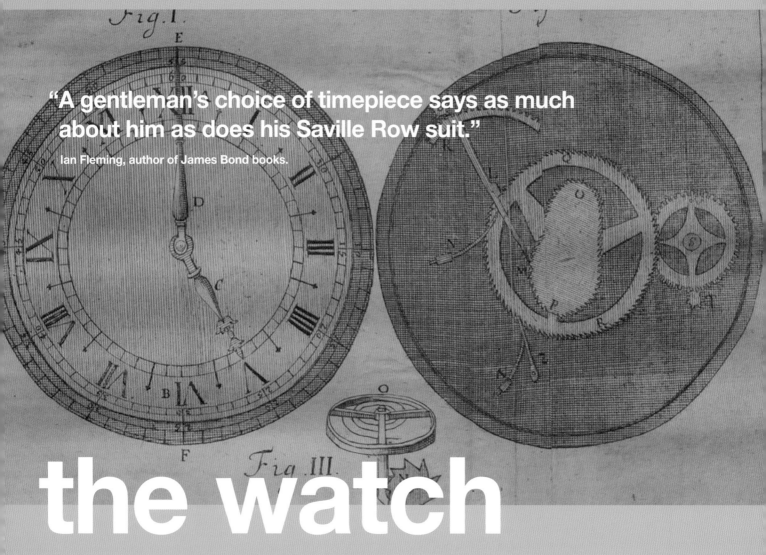

"A gentleman's choice of timepiece says as much about him as does his Saville Row suit."
Ian Fleming, author of James Bond books.

the watch

To some, it's the ultimate status symbol, to others it's just a simple, convenient timepiece, but either way, wristwatches have been a popular accessory for more than a century and despite the ubiquity of smartphones, look set to stay on many people's wrists for quite some time yet.

Clocks powered by water or rising and falling weights were common in Europe by the 15th and 16th centuries during which time, locksmiths and other craftsmen of intricate mechanisms began fashioning scaled down models as small as three inches long. A Nuremberg locksmith and clockmaker called Peter Heinlein is frequently cited as one of the key originators producing miniature clocks powered by a wound mainspring out of iron parts and frequently worn around the neck. Writing in 1511, Johann Cochläus stated that Heinlein, "still a young man, fashions works which even the most learned mathematicians admire. He shapes many-wheeled clocks out of small bits of iron, which run and chime the hours without weights for forty hours."

Early watches relied on parts called the wheeltrain and escapement to ensure that as the mainspring unwound, it transmitted its energy to the oscillating section of the watch called the balance. These first watches only featured an hour hand as the escapement and, indeed, the entire watch mechanism, was not accurate enough to depict minutes with any certainty. The development of a hairspring or balance spring that helped control the oscillations of the balance, attributed to both Englishman Robert Hooke and Dutch scientist Christiaan Huygens in 1670s.

The 19th Century saw the introduction of wristwatches, initially hand-crafted for noblewomen including what is widely-regarded as the first wristwatch in 1810 made by Europe's most celebrated watchmaker of the time, Abraham-Louis Breguet for Caroline Murat, Queen of Naples and sister of Napoleon Bonaparte. Some forward-thinking military men noted the wristwatch's practicality and potential. Germany's Kaiser Wilhelm I, for example,

Above An early spring-driven pocket watch drawing in the German scientific journal, Acta Eruditorum, 1737.
Right page Watches range in design and functionality with some possessing alarms, chronograph stopwatches, dual time and in the case of smart watches (center right) the ability to run apps.

No.60 the watch

Great Inventions We Take For Granted

ordered 2,000 Swiss wristwatches in 1880 for his naval officers for timing bombardments. But in general, men's watches stayed in their pockets and the wristwatch, or wristlet as it was known at the time, was mostly only spotted on women's arms until World War I. There, the rise of aviation, the timing of troop advances and the need for synchronous firing of large guns in planned artillery bombardments called for timepieces that could be seen immediately without fishing around within one's battledress. As more and more soldiers were equipped with wristwatches in active service, so the attitude to wristwatches for men changed back at home leading the Horological Journal to write, "the wristlet watch was little used by the sterner sex before the war, but now is seen on the wrist of nearly every man in uniform and of many men in civilian attire."

In 1926, German watchmaker Hans Wilsdorf announced the launch of the first fully waterproof wristwatch, the Rolex Oyster. By this time, the piezoelectric properties of quartz crystals had been discovered and in 1927, the first quartz clock was constructed. It took precisely four decades before the first quartz watch was announced in both Switzerland – the Beta 1 produced by the Centre Electronique Horloger, and Japan, the Astron by Seiko. Utilising the ability of quartz to vibrate consistently at a high rate, typically 32,768 times per second, these watches enabled highly accurate timekeeping with no gears or other moving parts, making reliable watches simple and highly affordable.

First standard waterproof watch
Oyster, Rolex, Switzerland - 1929

No.60 the watch

"Two decades earlier, men would rather have worn a skirt than a wristlet; after World War I, men wouldn't be caught dead without one...watches had become sexual status symbols for men everywhere."

Aja Raden, author of Stoned: Jewelry, Obsession, and How Desire Shapes the World

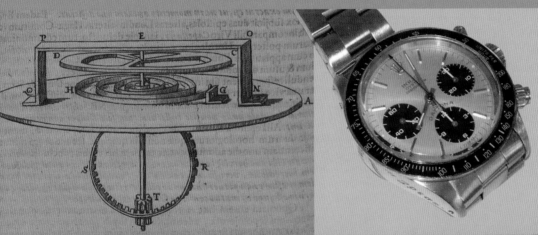

The Rolex Cosmograph Daytona worn by actor Paul Newman became the world's most expensive watch when it was sold at a 2017 auction for $17,752,500.

Left page A 1929 edition Rolex Oyster – the first waterproof and dustproof watch with its workings held in a hermetically sealed casing. **Above (top left)** Drawing of an early watch balance spring attached to a balance wheel, by Christiaan Huygens, published in his letter in the Journal des Sçavants in 1675. **Above (top right)** The Rolex Cosmograph Daytona originally introduced for racing car drivers in 1963. **Above** Wristwatches can convey style and status or be used to time runs and exercise periods.

Great Inventions We Take For Granted

Many countries, civilizations and empires deployed some form of postal system to move letters and other documents around their territory.

the postage stamp

Emperor Augustus established the cursus publicus, for example, in the Roman Empire almost 2,000 years ago. Most systems, though, were piecemeal and featured confusing series of charges. Many also required the receiver to pay for their mail and not the sender. The receiver could refuse to pay for their post and in some circumstances, the sender would write a short coded message or symbol on the outside of the envelope which the receiver could spot so they obtained the intended message without paying for their mail.

In the 1830s, the postal service in Britain was in a mess and reformer Rowland Hill was keen to do something about it. He suggested a single nationwide charge for letters weighing half an ounce or less of one penny which would be paid for by the sender and with a small stamp which would act as proof of payment. Hill suggested that the stamp should be, "a bit of paper just large enough to bear the stamp, and covered at the back with a glutinous wash." In May 1840, the world's first national postage stamp went on sale in Britain – known as the penny black for its colour and cost. More than 68 million of these stamps bearing a portrait of Queen Victoria were sold.

The US Postal Service had been instituted in 1775 with Benjamin Franklin its first Postmaster General. Weeks after word of the penny black reached American shores in 1840, Senator Daniel Webster of Massachusetts introduced a resolution in the US Senate to reduce postal rates and for to follow Britain's lead and issue national prepaid postage stamps. Following an act of Congress, the USPS issued its first national postage stamps which went on sale in New York City in July 1847. Customers were offered two rates – a 10 cents stamp featuring George Washington and a 5 cent stamp featuring Ben Franklin – the first non-head of state on any postage stamp in the world. In 2018, 16.5 billion U.S. postage stamps were printed. The first self-adhesive stamps, initially designed to counter the humid environments of countries with

No.61 the postage stamp

Left page (left) A stamp printed in the USA showing Statue of Liberty, circa 1941. **Left page (right)** A stamp printed in the USA showing 21c series, circa 1973.
This page (above) Old postage stamps from Italy.
This page (far left) The iconic Penny Black stamp was sold from May 1840 to February 1841. A mint condition example of this stamp can fetch over $5000. **Left** Old postcards with postage stamps.
Below British Guiana stamp issued in 1856.

tropical climates which could cause unused stamps to stick together, were issued by the African nation of Sierra Leone in 1964.

Strange Stamps
Most stamps are small, made of paper and rectangular. Some postal services have occasionally pushed the boundaries of stamp design. These include the banana shaped design produced by Tonga in 1969, a stamp made not of paper but cork by Portugal in 2007 and three years earlier an A 7.5 Euros stamp issued in Austria which was covered in Swarovski crystals. A set of plastic postage stamps issued by Bhutan in 1972 acted as phonograph records and could be played on a record deck at 33 1/3 RPM. In 2018, the US issued a series of Frozen Treats stamps – the first American 'scratch and sniff' stamp which released summertime scents when scratched.

US Stamp Firsts
The first woman to appear on a U.S. postage stamp was Queen Isabella in 1893 who appeared on a $4 stamp to commemorate Christopher Columbus' voyages of discovery. Isabella was the Queen of Castille who sponsored Columbus' expeditions. The first American woman honored on stamp was Martha Washington in 1902. The first Native American featured on a stamp was Pocahontas in 1907 whilst the first African-American to feature on a stamp was Booker T. Washington in 1940.

In 2014, Sotheby's in New York City auctioned a British Guiana One-Cent Black on Magenta stamp for $9.5 million to an anonymous buyer – the most paid for a single stamp.

play

Great Inventions We Take For Granted

the bikini

Some inventions are completely new, and some out of the blue. Others are rediscoveries and re-imaginings of earlier innovations. In post-war Europe, two French fashion designers battled for the title, inventor of the modern bikini, but this two piece bathing suit has a far longer history than might be first realised.

Active women such as dancers in ancient Greece wore a breastband of material called a mastodeton or an apodesmos whilst mosaics dating from around 300CE from the Villa Romana del Casale in Sicily were discovered by Italian archaeologist, Gino Vinicio Gentili during excavations in 1959 and 1960. One scene, labelled, Coronation of the Winner, depicts ten women exercising and playing sports including running, discus-throwing and weightlifting in something close to a modern approximation of a bikini.

Another relic of ancient Rome was discovered in Pompeii in the late 18th Century during the first investigations into the city buried by a devastating volcanic eruption of Mount Vesuvius in 79CE. A statue of the Greek goddess of beauty, Aphrodite taking off a sandal but wearing a gold bikini, was deemed so shocking that it joined a cache of other sexually explicit finds from Pompeii and elsewhere in the Gabinetto Segreto, secret collections in Naples, only viewable by men of important social standing who received permission from the authorities.

An extremely prudish attitude to the female body and bathing continued throughout much of the western world during the 19th and early 20th centuries. Women would cover themselves head-to-ankles in clothing, often made up of multiple layers, before they could even contemplate bathing in public. When famous Australian swimming champion Annette Kellerman went to swim at Boston's Revere Beach in 1907 she was arrested and tried in court for indecency because her one-piece outfit whilst covering all of her upper body ended above the knees.

Designers in Europe began producing modest long two piece bathing suits which gradually shrunk in size revealing a thin strip of midriff in the early-to-mid 20th Century. Jacques Heim produced the, for the time, daring Atome bathing suit for young women in 1932, but re-launched it after World War II when he owned sportswear lines and a

No.62 the bikini

This page (bottom) Villa Romana del Casale, Sicily. UNESCO World Heritage Site. **Below** A beach vendor selling bikinis carries his merchandise along Copacabana beach. **This page (top left)** A model poses in a 1950s gym wearing a floral two-piece swimsuit. **Left** A model walks the runway for Gigi C Bikinis during the Paraiso Fashion Fair. **Right page** Women sporting a variety of swimsuits including two-piece suits wave for the cameras in the 1950s.

beach shop in Cannes on the French Riviera. It was advertised as the smallest swimsuit in the world both in print and by skywriting planes flying above the beaches of southern France.

Louis Réard was a French mechanical engineer who was running his mother's Parisian lingerie business. He went far further than Heim, making a bathing suit which exposed the navel when he constructed the first string bikini consisting of four triangles of cloth in a faux newspaper print design tied together by narrow cords. Just 30 square inches of material were used to make each of his prototypes which no professional fashion model would wear so he demonstrated his design employing a nude dancer from the Casino de Paris called Micheline Bernardini. The swimwear was launched four days after the United States' first nuclear test explosion at Bikini Atoll in the Marshall Islands, giving Réard a modern instant name for his outfit which stuck. He also employed skywriting planes to trump Heim by stating that his outfit was, "smaller than the smallest bathing suit in the world." In later years, he would state that no swimwear should be considered a bikini unless it could be pulled through a wedding ring.

Early sales were slow and with many critics, the bikini was initially banned from many beauty pageants and outlawed completely in a number of western countries including Spain, Belgium, Portugal, Italy and Australia. The tide turned in the late 1950s and 1960s, partly due to the impact of surf movies and bikini-wearing actresses including Brigitte Bardot, Raquel Welch and Ursula Andress in the 1962 James Bond movie, Dr. No. Two years later, Sports Illustrated magazine debuted their swimsuit edition with a white bikini on its front cover.

> "It is hardly necessary to waste words over the so-called bikini since it is inconceivable that any girl with tact and decency would ever wear such a thing."
> Modern Girl Magazine, published in the US in 1957.

143

Great Inventions We Take For Granted

> "The flush toilet may have been the most civilised invention ever devised, but the remote control is the next most important. It's almost as important as sex."
>
> Eugene Polley, in a 2002 interview.

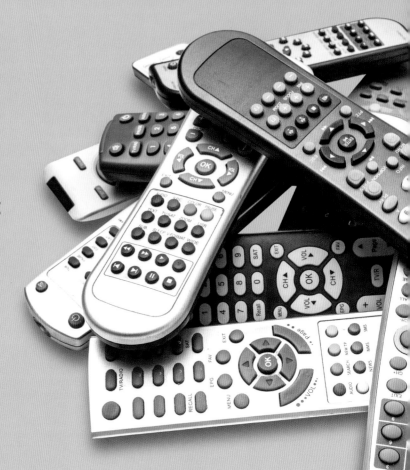

the tv remote

Zenith Radio Corporation's founder, Eugene F. McDonald, famously hated commercials and dreamt of a time when TV advertising would collapse and disappear. In the meantime, 'The Commander' as he was nicknamed, instructed his engineers to give Zenith's burgeoning range of consumer television sets some form of remote control, allowing him and others, to change channels when a commercial break appeared without having to shift more than a finger muscle.

The first TV remote control was Zenith's Lazy Bones in 1950, breathlessly advertised in newspapers as, "It's like something out of Arabian Nights." This remote control allowed channel hopping from the comfort of the couch by pressing a handheld button. The controller only worked with selected new Zenith TVs to which it was connected to by a long cable that trailed across a living room. Consumers reported tripping up on the cable and a wireless solution was called for. Zenith engineer, Eugene B. Polley, developed the Flash-Matic in 1955 for which he received a $1000 company bonus.

Coloured green and shaped like a space age ray gun, the Flash-Matic was essentially a flashlight that viewers shone at a TV set specially equipped with four photocells behind window openings built into the front corners of the set acting as light sensors. Viewers pointed the gun at one of these four photocells and pulled the trigger to send a beam of light, for example, to the bottom right window if they wanted to toggle the sound either on or off or the top left window if they wanted to turn the television's tuner dial anti-clockwise. It cost a hefty $100 (more than the average weekly wage of the time) and could be confused by strong sunlight streaming in through a window.

Robert Adler developed Zenith's third remote control system, known as Space Command. This large box contained four chunky buttons – one to power the television on or off, two tuning buttons and a volume mute control. A button press caused a small hammer inside to strike a two and a half inch long aluminium rod which produced a high frequency tone inaudible to human hearing but registered by the television. Like the Flash-Matic, Adler's remote wasn't perfect and could sometimes be set off by jangling keys or rattling coins but the principle of using ultrasonic signals to control a TV set would remain the industry standard for a quarter of a century during which time Zenith, alone, sold more than

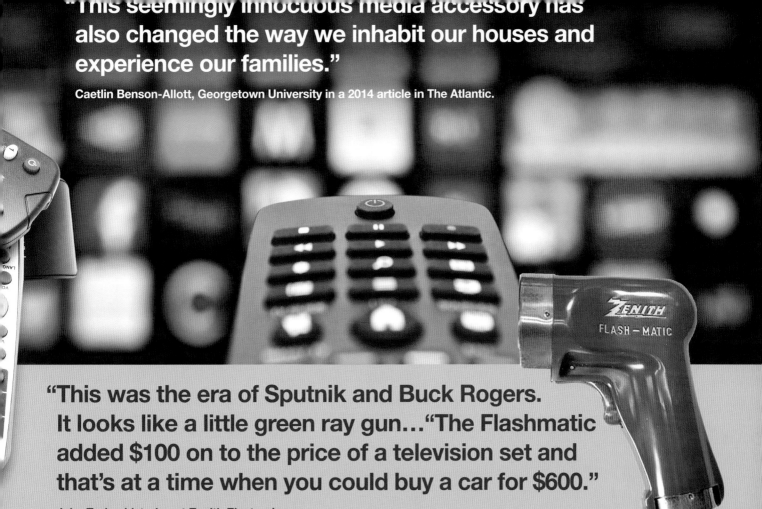

"This seemingly innocuous media accessory has also changed the way we inhabit our houses and experience our families."

Caetlin Benson-Allott, Georgetown University in a 2014 article in The Atlantic.

"This was the era of Sputnik and Buck Rogers. It looks like a little green ray gun…"The Flashmatic added $100 on to the price of a television set and that's at a time when you could buy a car for $600."

John Taylor, historian at Zenith Electronics.

nine million TV sets controlled by Space Command remotes.

Remotes adopting infra-red and RF (radio frequency) communications gradually replaced ultrasound in the 1970s and 1980s, but otherwise, the basic form and function of the long rectangular 'candy bar' TV remote has changed little, only its complexity as TVs grew smarter and their attributes, from picture size to subtitling, more controllable by the user from their sofa. These remotes, though, are likely to gradually fade away into disuse as smartphone and digital assistant apps replace button pressing with voice-activation and gesture recognition – a potential boon for those millions of Americans who spend time each week hunting around the couch and living room for a lost remote.

Steve Wozniak

Apple co-founder Steve Wozniak left the company to invent one of the very first universal remote controls, the 20-button CL9 CORE in 1987.

Top left A pile of remotes, similar to those that litter many people's homes. Lost remotes are one of the bugbears of modern life. **This page (top)** A remote points at a smart TV capable of selecting apps or options at a touch of a button. **Above (right)** Zenith's Flash-Matic exhibits its space age design. A working model in good condition is a collector's item today.

Great Inventions We Take For Granted

From gridiron to soccer, you cannot attend many team sports matches without your ears registering the sound of the referee, umpire or another official's whistle.

the whistle

They're also found on most passenger airliner lifejackets, blasting on the beach as a lifesaver broadcasts a warning to bathers or noisily accompanying drums and other instruments at carnivals, festivals and parades.

Whistles made of bone or wood have been used for thousands of years in religious ceremonies and for practical purposes to attract attention. Boatswain's pipes or whistles are a direct descendent of whistles used on ancient Greek and Roman oar-powered naval ships and used to keep the rhythm of the oarsmen's stroke. Historians believe the whistle extends even further back, as far as 4,000-5000 years ago in China when used to alert a settlement to an incoming hostile invasion.

Joseph Hudson, a toolmaker from Birmingham, England, revolutionised whistle design from the 1870s when he converted his small washroom into a workshop. A Hudson brass whistle was used in a soccer match for the first time in 1878 during a second round Football Association (FA) Cup game between Nottingham Forest and Sheffield. Before that point, signals where communicated by the umpires (this was before a single referee was placed in charge of a soccer match) to the players by shouting, brandishing a stick or even waving a handkerchief.

In the 1880s, Hudson decided to add a small ball or pea inside his design of whistle to help produce a distinctive and ear-catching trilling sound. The air stream from a person blowing the whistle's mouthpiece is split by the bevelled opening. The remaining air inside the whistle chamber propels the pea round the chamber. As the ball rises and falls in the chamber via air turbulence and blocks off part of the sound hole, it produces variations in the pitch of the whistle's sound.

Hudson introduced his pea whistle, the Acme Thunderer in 1884. It remains the world's best-selling whistle with over 200 million purchased and has been used by referees at FIFA World Cup soccer tournaments as well as by officials at Olympic games. Hudson also equipped London's Metropolitan Police with whistles which could be heard over a distance of a mile as well as armed forces which signalled manoeuvres with a shrill blast during both World War I and World War II.

Most whistles followed Hudson's lead and were made of brass, often plated with nickel or chromium until 1914 when the first manufactured plastic whistles were produced, in Britain. Earlier attempts to produce a satisfactory model from vulcanite (hardened rubber) had failed. Modern plastic whistles are either glued or use ultrasound to weld the plastic parts together.

The Fox 40
Canadian basketball referee, Ron Foxcroft designed a new pea-less whistle which first went on sale in 1987. It is tuned to produce three slightly different sound frequencies simultaneously to produce a piercing vibrato sound that cuts through the air. The Fox 40 was debuted at the 1987 Pan American Games in Indianapolis, Indiana and quickly found favour with police forces, the US Coastguard and sports officials. It has since become the official whistle of the National Hockey League and the NBA.

Whistles are made from a wide range of materials including brass **(top left)**, plastic **(second from top left)** and carved wood. **(second from top right)**. They are used for an equally wide range of tasks by sports officials, law enforcement officers, teachers and revellers at festivals.

146

No.64 the whistle

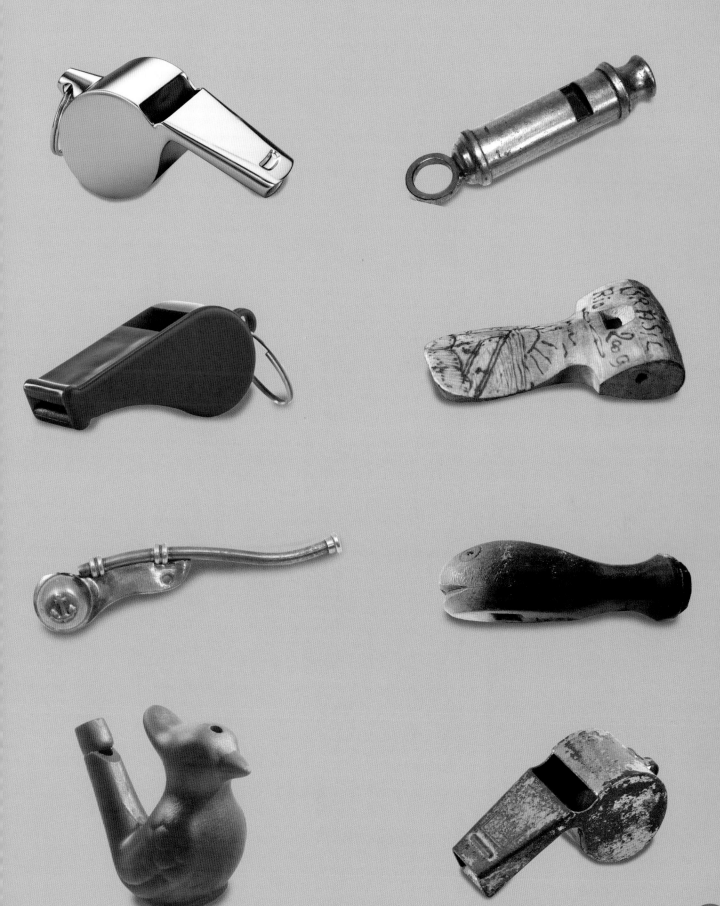

Great Inventions We Take For Granted

the swiss army knife

Few travellers, anglers or outdoorsmen and women leave home without a trusty pocket knife packed full of different blades and tools.

They stem back to a 120 year old desire to equip the soldiers of Switzerland with a portable tool. To this day, each military recruit (which includes the vast majority of the male population serving their compulsory military service) receives a Swiss Army knife as they begin their military service.

The very first commission for 15,000 knives could not be fulfilled by a company in Switzerland so German knife maker, Wester & Co supplied the order in 1891 with a Modell 1890 knife which contained a screwdriver, can opener and a single blade. Seven years earlier, Karl Elsener along with his mother, Victoria, had opened a handcrafted knives and cutlery business in the Swiss canton of Schwyz. They developed a foldable penknife that could be used by soldiers to open tins, cut cord and assemble and disassemble their rifles. After trialling it with Swiss army units and making changes to the design, the Elsener family began supplying portable, multi-purpose knives to the military. In the 1890s, Karl Elsener made further improvements, working out that he could put blades and tools on both sides of the handle using the same spring to hold both sides in place. More implements could be fitted into a knife, increasing its versatility and usefulness.

From 1908-09 onwards, the company had to split Swiss Army orders with a rival company, Wenger, to avoid regional favouritism. Both companies' knives weren't well-known outside of central Europe until World War II when British and American forces came across them, and struggling with their original name, Schweizer Offiziersmesser, dubbed them "Swiss army knives." More than 400 different models have been constructed since including specialist models aimed at fishermen, golfers or cigar smokers and from small keyring models up to the Wenger Giant – the world's largest production Swiss Army knife. Weighing in at 32 ounces and measuring nine inches wide, this behemoth was launched in 2006 and contains 87 implements including a flashlight, tread gauge, rivet setter and a laser pointer, all in all, offering 141 functions.

Top (left) A large Swiss army knife with all its blades closed remains a compact object easily packable in any bag or pocket. **Center** Opened up, the knife offers a wide range of blades and tools from pliers, a saw and scissors to bottle openers, screwdrivers and mini tweezers stored in the outer casing.

No.65 the swiss army knife

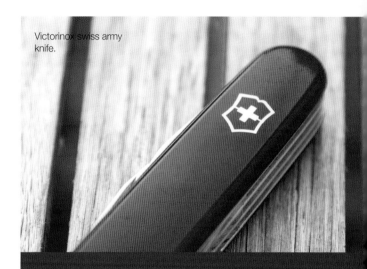

Victorinox swiss army knife.

In 1978, NASA placed an order for 50 Swiss Army knives – the Master Craftsmen model. These were wielded by astronauts both in simulations on Earth and in space missions and NASA made a further order of knives in 1986 to supply Space Shuttle astronauts. In 1995 Canadian astronaut Chris Hadfield on a Space Shuttle mission to the Russian space station, Mir, used his Master Craftsman knife on an EVA to cut through all the tape that sealed a door so that the Shuttle could successfully dock with the space station.

The company was only named Victorinox after the death of Elsener's mother in 1909. The "inox" in the name comes from a French term for stainless steel. Four generations of the Elsener family have been in charge of Victorinox: Elsener's son, grandson and great-grandson, all named Carl. The company manufactures approximately 34,000 knives every day.

Great Inventions We Take For Granted

the hammock

The ultimate vacation or lazy day destination, hammocks offer a relaxing respite from a hectic world.

Hammocks are thought to have originated in Central and South America over 1,000 years ago. They may have developed from ropes covered in cloth or plant matter or grown out of reinforced fishing or hunting nets strung between two trees or supports. The ancient Maya made extensive use of these suspended beds woven from sisal or other tree bark. Use spread throughout the Americas prompting Portuguese-born historian, Pero de Magalhães Gandavo, to write in 1570, "Most of the beds in Brazil are hammocks, hung in the house from two cords."

The Taino people in the Bahamas suspended their netted hammocks above the ground with a small fire or raked glowing embers underneath for a supply of smoke to keep snakes as well as mosquitoes and other harmful insects at bay. The first European to visit the Caribbean, Christopher Columbus, reached the Bahamas in 1492, where he discovered the Taino and how they slept in hamacas, samples of which were taken back to Europe where initially they were little more than a curiosity. The name would eventually be anglicized to hammock.

The ease with which hammocks could be manufactured and strung between posts or hooks, saw them adopted below decks by European sailors in the late 16th Century onwards. Sailors enjoyed a more comfortable night, swayed to sleep on the ocean's swell and appreciated being raised above the wet and often filthy and rat-infested deck. The navies of Europe preferring a heavy-duty version made from stout canvas whilst woven netting hammocks remained common on the other side of the Atlantic. Many ships connecting towns in South America still make use of net hammocks for passengers to rest in, but the hammock has found favour elsewhere. As a space-saver, it is part of many backpackers essential kit, especially in tropical wilderness areas where it can be more easily enclosed in a mosquito net than other bed types. Long distance truck drivers on the Indian subcontinent lacking a sleeping cab even string hammocks up underneath their vehicles to take a well-earned rest.

Above A romantic liason occurs on a hammock decorated with woven patterns and fringing. **Right** Hammocks hanging below decks on the historic ship, USS Constellation moored in Baltimore Harbor. **Right page (top)** A cyclist rests in a forest after stringing up his lightweight hammock between two trees. **Far right** Close Up of Fabric of Brightly Colored Woven Hanging Hammocks in a Street Market.

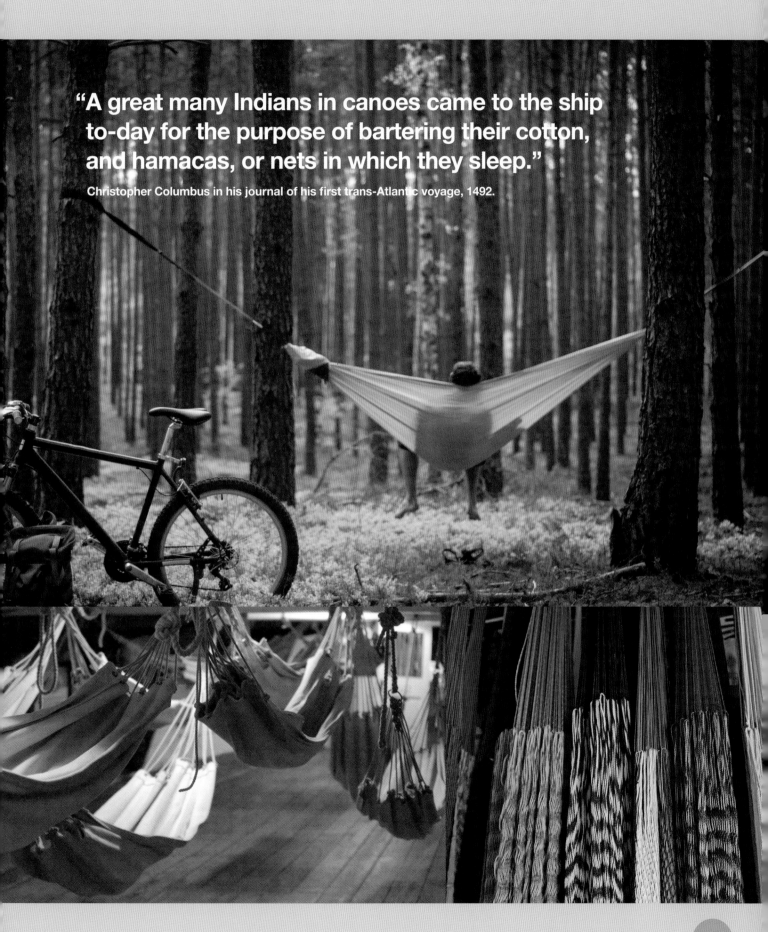

No.66 the hammock

"A great many Indians in canoes came to the ship to-day for the purpose of bartering their cotton, and hamacas, or nets in which they sleep."

Christopher Columbus in his journal of his first trans-Atlantic voyage, 1492.

Great Inventions We Take For Granted

"Because it was so new, people would invite the whole neighborhood to come hear it. They wanted to run their own hi-fi show. That's kind of the way it worked all along. You had to hear it to believe it."

John C. Koss on the invention of consumer stereo headphones.

headphones

"Headphones give us absolute control over our audio-environment, allowing us to privatize our public spaces."

Derek Thompson, The Atlantic, 2012

No.67 headphones

Introduced at the 2017 Consumer Electronics Show, Tournaire's Focal Utopia headphones, encrusted with six carat diamonds, are the world's most expensive at a cool $120,000 per pair.

Above (from left to right) A pair of 1970s circumaural headphones fitted snugly around the ears to surround the listener in sound. On ear headphones with curled leads were popular in the 1980s **(top)**. Old headphones with a lightly padded headband but no padding on the earpieces **(bottom)**. **Above** Bluetooth earbuds offer wireless listening providing a Bluetooth source is close by. **Above right** A child listens to music from their smartphone using wired headphones.

Whether in-ear or on-ear, headphones formed a crucial part of personal sound systems that enabled people for the first time in history to listen to recorded music on the move.

Headphones developed out of headsets used by early telephone switchboard operators. These listening devices frequently contained one speaker and could weigh as much as 11 pounds making them uncomfortable to wear around the head and neck for long periods. One of the first instances of such headsets reaching the home was the Compagnie du Théâtrophone service in France in 1890, closely followed by the Electrophone service in Britain. Users paid an annual subscription fee (£5 for the Electrophone which was more than three weeks wages for a typical labourer of the time) and could phone up and listen using the headset to a live performance from a theatre or opera.

Fundamentalist Mormon, Nathaniel Baldwin invented the first pair of comfortable audio headphones in 1910 – effectively binding the loudspeakers from two telephone receivers to an arching band worn over the head. Impressed by their comfort and sound quality, the US Navy ordered 100 pairs – an order which Baldwin fulfilled by building them on his kitchen table in Utah. Sound quality improved in the 1920s and 1930s with dynamic headphones and with the arrival of stereo headphones in 1958. Milwaukee jazz musician and businessman John C. Koss had been looking for commercial opportunities to follow his hospital television business. Teaming up with engineer, Martin Lange Jr, the pair developed a small, affordable record player and a pair of stereo headphones, the Koss SP/3, to go with it.

Their timing was fortuitous with the flourishing of popular music, a growing teen passion for rock n roll and a resulting boom in radio channels and record sales. Millions of American homes bought headphones so that members of the family could listen to music without disturbing the rest of the household. Long before Beats by Dr. Dre, Koss even produced the first pair of celebrity endorsed 'cans' the Beatlephones in 1966. Most models bought throughout the 1960s and 1970s were circumaural or closed headphones fitting over the entire ear, but the invention of the Sony Walkman personal cassette player in 1979 and, later, portable compact disc players saw an increase in supra-aural or on-ear headphones which used foam pads to nestle on the ears and admitted some degree of environmental sound in. Small earbuds and canal phones that fit into the ear canal, only became popular in the late 1990s and onwards and were given a boost by the arrival of iPods, other solid state digital music players and, later, smartphones.

"Beauty, to me, is about being comfortable in your own skin. That, or a kick-ass red lipstick."

Gwyneth Paltrow, actress, quoted in Goddess by Elisabeth Wilson.

lipstick

Cosmetics are a far from modern invention. Both ancient Egyptian women and men, for example, wore elaborate eye make-up featuring green malachite paint and thick black kohl eyeliner made from fat, soot and crushed up minerals such as galena. Ancient Greeks sometime reddened their cheeks with beetroot whilst some ancient Romans used nail polish concocted from fats mixed with sheep's blood.

Mesopotamian women around 3000BCE were the first known peoples to use lip colours. Archaeologists have discovered artefacts depicting women applying a painted paste made of beeswax mixed with ground up semi-precious gemstones. Women in other ancient cultures applied differing concoctions with a brush or their fingers, some containing crushed berries or red ochre rock whilst fish scales were often added to the mixture to give lips a glossy sheen. Ancient Egyptian queen, Cleopatra VII was said to have applied lipsticks using the waxy eggs of ant larvae as a base mixed with the vibrant red of crushed beetle species.

In Medieval Europe, the reddening of a woman's lips was seen as a challenge to god and often associated with heresy and witchcraft. It is a myth that the British parliament banned lipstick in 1770 as a devilish attempt to trick men into marriage, but even so lipstick remained socially unacceptable in many societies deep into the 19th Century.

In 1870, the French perfume house, Guerlain introduced the first modern lipstick for sale, named, "Ne m'oubliez pas" (Forget me not). It consisted of a waxy base containing castor oil, beeswax and deer tallow coloured using carminic acid obtained from the cochineal – a small scale insect found in Mexico and Central America. Unlike most modern lipsticks, early lipsticks were sold in a block or in a silk bag and applied with a brush. Debate surrounds the identity of the inventor of the classic lipstick tube. The Scovil Manufacturing Company in Connecticut may have produced metal tubes with a small lever to advance the lipstick upwards as early as 1915 whilst push-up lipstick tubes may have existed in Europe a few years earlier. What is certain is that the first known patent for the classic swivel up lipstick tube was applied for in 1922 and granted the following year to Nashville, Tennessee resident, James Bruce Mason Jr.

Smudge-Free Smooch
A formula for the first smudge proof lipstick was developed by Hazel Bishop's tireless experimentation over many years in her kitchen away from her day job as an organic chemist (she had, for example, helped develop aviation fuels for Standard Oil during World War II). Bishop found that using lanolin as a moisturiser helped offset the bromo acid dies which produced a long-lasting lip coating at the cost of lip dryness. Launched in 1950 with the advertising slogan, "it stays on YOU, not HIM!" Bishop's No-Smear Lipsticks retailed at one dollar each and captured a sizeable market share within months.

This page (top left) A Japanese geisha girl exhibits the striking red colour of traditional lipstick originally called Komachi Beni and applied with a brush. **This page (center)** The allure of dark red lipstick remains strong even with shifting fashions over the decades. **Right page (bottom)** A small selection of the vast range of colours lipsticks typically are available in. **Right (top)** A model wears a striking luminous green lip colour.

No. 68 lipstick

National Lipstick Day is held every July 29th.

155

Great Inventions We Take For Granted

the newspaper

The arrival of the movable type printing press in Europe, pioneered by Strasbourg's Johannes Gutenberg, saw books, pamphlets, handbills and other printed materials begin to spread through that continent at a rate simply not possible before when most texts were still copied by hand or woodblock printed – both slow, laborious processes. New information and ideas about politics, science and society started to be communicated and disseminated more quickly and widely than before, prompting a thirst for knowledge amongst more people, and this included a desire to gain the latest news in written form.

Venice quickly became a major printing centre and in 1556, its government started to produce a weekly newsletter, nicknamed a gazzetta or gazette, after the name of the low value coin it cost to purchase. The first regular newspaper was printed by Johann Carolus back in Strasbourg from 1605 onwards. This former bookbinder and bookseller had previously hand-copied out the news he received but purchased a printing press to produce a higher circulation of his Relation aller Fürnemmen und gedenckwürdigen Historien (Account of all distinguished and commemorable news). This publication was recognised in 2005 by the World Association of Newspapers (WAN) as the planet's first printed newspaper.

Newspapers began to spring up throughout Europe with Amsterdam becoming an early centre of activity and the location of the first English language news sheet, the Corrant out of Italy, Germany, etc. in 1620 whilst the first successful daily newspaper in Britain, The Daily Courant, was first printed on 11 March 1702 by E. Mallett in London. It was designed initially as a single page news sheet summarizing the news from other newspapers from the continent. The paper lasted until 1735 when it was merged with The Daily Gazetteer whilst the location of Mallett's printing premises, on Fleet Street in London, would become the centre of the UK's national newspaper trade for 250 years.

Whilst single page papers, known as broadsides, were published under colonial control in America in the 17th Century, the first attempt at a multi-page newspaper, Publick Occurrences, Both Foreign and Domestick lasted just one issue in September 1690 before Benjamin Harris' pioneering paper was suppressed by the authorities. The thirst for news, though, took

Above (from left to right) Vintage printing press plates. Boys selling early single-page newspapers accost an American gentleman, illustration originally published in 1880.

No.69 the newspaper

Above A stack of freshly printed daily newspapers transported to a printing plant. **Right** The first ever New York Times front page, 1851. **Above (right)** A newsboy in Pennsylvania sells papers on the street, 1910. **Below (right)** French people flock to an illuminated newspaper kiosk after dusk, 1857.

hold and a series of news sheets and weekly newspapers were published before America gained independence. In 1783, the Pennsylvania Evening Post became the first American daily. As a weekly back in 1776, the July 6th issue became the first to publish the newly-adopted Declaration of Independence.

Back in Europe, newspaper printing gained a major boost in 1812 with the invention by German engineers, Andreas Bauer and Friedrich Koenig, of a new high speed printing press powered by a steam engine. Five times faster than previous mechanical presses, it was first used to print The Times newspaper in London in 1814 and helped usher in an explosion in newspapers and increasing circulations.

Great Inventions We Take For Granted

The little black dress has become such a popular staple that the abbreviation of its title, LBD, entered the official Oxford English Dictionary in 2010.

the little black dress

The essential outfit in every woman's wardrobe, a little black dress offers the perfect backdrop for mix-and-match accessories to transform this simple item into anything from a wild, head-turning party outfit to a sensible, formal attire.

Whilst black had been worn by women in the distant past, entering the 20th Century it was most closely associated with servants in some societies (who might wear black dresses and white aprons and hats) and, of course, was the colour of mourning. As the 'roaring twenties' progressed, many designers concentrated on producing colorful and embellished clothes in the 'flapper' style to match the prevailing upbeat mood. Gabrielle 'Coco' Chanel chose a different path and in 1926, her design for a simple calf length black dress in crêpe de Chine. with long narrow sleeves and accessorised only by a long string of pearls was published in Vogue. The magazine, with considerable foresight, stated that the design would become "Chanel's Ford" referring to Henry Ford's Model T – the standard for affordable, reliable cars at the time and that it could become, "a sort of uniform for all women of taste."

Whilst other designers had also produced black or dark colored dresses with simple lines, it was Chanel that became associated with the design, as she produced further versions around the same theme including a sheer black lace dress with a matching capelet that was profiled in Vogue in 1930. Chanel later said, "I imposed black; it's still going strong today, for black wipes out everything else around."

The little black dress was popularised by starlets in the 1950s and 1960s and received arguably its biggest boost when Audrey Hepburn played the role of Holly Golightly in the smash hit move, Breakfast at Tiffany's. Penny Goldstone, writing in Marie Claire magazine in 2017 stated, "There's no arguing it's one of the most famous dresses in cinematic history. Come on, is there a more iconic scene than the one of Audrey munching on a croissant and sipping coffee while admiring Tiffany's jewellery, all the while wearing her Givenchy dress from the previous night's party?" One of the original dresses intended for the film made by Hubert Givenchy for Hepburn was auctioned in December 2006 at Christies in London for a £467,200 (approximately $900,000).

No.70 the little black dress

"One is never over-dressed or under-dressed with a Little Black Dress."
Karl Lagerfeld

Great Inventions We Take For Granted

> "Unlike other mortals, the 42,000 members of the Diners' Club need never pay the waiter when they wind up a spirited evening on the town. They simply sign the check, get billed once a month."
>
> TIME magazine, 1951

the credit card

Those wallet and purse-friendly slivers of plastic have proven indispensable to many and are increasingly reducing reliance on cold hard cash in people's pockets.

Such is the ubiquity of paying by plastic in varying ways – from charge cards to credit and debit cards - it is hard to remember how short a time it was when cash was king and credit or charge cards a rarity or option. Whilst some US department stores introduced proprietary cards giving customers a limited form of credit in the first quarter of the 20th Century, the story of paying by plastic can be traced back to 1946 when Brooklyn banker, John Biggins launched the Charg-It card. Purchases made with the Charg-It card were forwarded to Biggins' bank, the Flatbush National Bank, which acted as the middleman, charging the customer and reimbursing the seller. Charg-It was only a localized scheme operating over a few blocks of Brooklyn for the benefit of the bank's customers, but the next scheme was both national and influential. According to legend, executive Frank McNamara was in a restaurant at the end of a business dinner when he realized with horror that he had forgotten his wallet. The embarrassment of waiting until his wife arrived with the necessary cash caused him to think up new ways of paying that didn't rely on cash. In reality, McNamara worked for finance company, Hamilton Credit Corporation who were looking for new business and the Diner's Club card the launched which could be used in a range of restaurants but charged interest on monthly payments billed to users proved something of a cash cow. Thousands signed up for a card and within five years, a raft of similar cards were available including Gourmet Club, Trip Charge, Golden Key and Carte Blanche. In 1958 both American Express and the Bank of America entered the fray with general purpose cards that could be used for more than travel and dining. American Express replaced paper cards with plastic in 1959 with Diners Club following two years later. By the 1970s, these plastic cards featured technology developed by IBM, to store key data on the card in the form of a small magnetic strip that could be "read" when placed in a pay terminal. Magnetic strips were, in turn, replaced by a small embedded microchip which communicates in encrypted code with the terminal to obtain authorization for the payment to be made. In recent times, many nations have introduced contactless payment using radio-frequency identification (RFID) or near field communication (NFC)-enabled microchips on cards, permitting mostly small transactions with a simple tap of the card on a terminal.

No.71 the credit card

"There will be only two classes of people - those with credit cards and those who can't get them."

Alfred Bloomingdale, department store magnate, making a prediction in 1960.

Top (left to right) A consumer makes a transaction using a portable card payment machine, Online ordering from home over the Internet using a laptop and a credit card, A wallet packed with different credit and debit cards. **Above (left to right)** A transaction is made with a wave of a card close to a contactless payment console, Credit card companies like Mastercard, American Express and Visa are global players offering vast networks of transaction possibilities.

In 2017, according to the Federal Reserve, US consumers made a staggering 6.6 trillion dollars of payments on credit, debit, and prepaid cards.

Great Inventions We Take For Granted

Whether sneaking in a quick blast of Candy Crush or Clash of Clans on a smartphone or settling down for a serious immersive session of Call of Duty, Fortnite or Grand Theft Auto, computer gaming now occupies a significant slice of children and adults' leisure time.

computer games

The 1950s saw the very first computer games running on giant mainframe computers, often written and programmed as tests of the machine's capabilities or an exploration of programming art and skill. OXO, a simple tic-tac-toe game for example, was developed by British computer scientist, A.S. Douglas in 1952 for the EDSAC computer at Cambridge University as part of his doctoral thesis on human-computer interaction. A little earlier, in late 1951, Christopher Strachey produced a checkers game for the Manchester Mark I computer housed in the northwest of England whilst in 1954, researchers at Los Alamos, home of atomic bomb development, programmed an IBM 701 computer to play blackjack. Another IBM 701, running Arthur Samuel's checkers program debuted on American television in 1956.

Early computing lacked the processing power, speed and display capabilities to create true moving graphic action games. The arrival of a DEC PDP-1 at the Massachusetts Institute of Technology in 1961 prompted a group of ex-students and research assistants, including Steve Russell, Peter Samson and Dan Edwards, to develop the first action game which they named SpaceWar! It pitted two players, each controlling and moving a spaceship and firing missiles at their opponent whilst avoiding the gravitational pull of a star. Given the limitations of the computer the game was written on, it was a remarkable achievement and versions of SpaceWar! were transferred or programmed on computers in a number of research centres all over the United States.

Nolan Bushnell encountered SpaceWar! in the late 1960s as a student at the University of Utah. In 1970, he began work developing an arcade game version of the game with Ted Dabney on behalf of Nutting Associates who sold it from 1971 onwards as Computer Space, the first mass-produced arcade computer game with some 1,500 machines built. Bushnell and Dabney formed Atari with a miniscule budget of $500 in 1972. The company produced their first arcade machine, running the tennis game, Pong by the end of the year using a TV bought from a store, a coin mechanism from a Laundromat and a milk carton inside to catch the coins. It was deployed in a bar in Sunnyvale, California bar and such was its popularity that the machine broke down when its coin catcher became overfilled with quarters. Pong proved a major success with some 30,000 machines manufactured at a time when 2,000 or more sales were deemed a success.

No.72 computer games

"The simple, classic games, where we didn't have those graphics to fall back on, had to be really well-tuned, and the response times had to be honed. We focused more on gameplay than I think people do today."

Nolan Bushnell, The Guardian newspaper, 2009

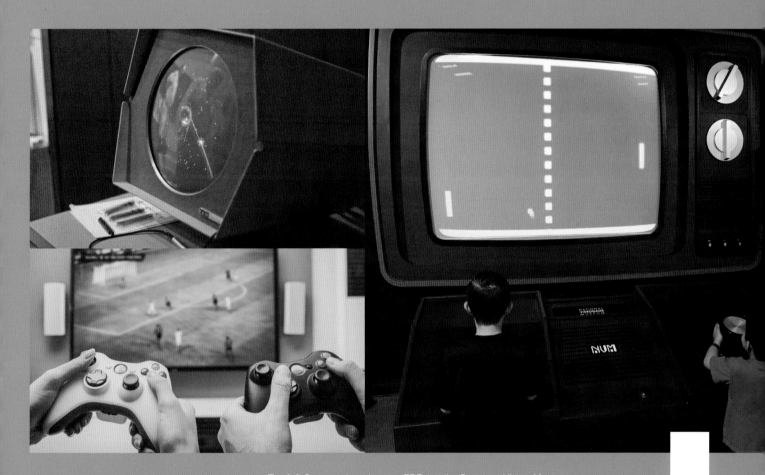

Top left Spacewar! running on a PDP-1 at the Computer History Museum, California. **Above (left)** Two players use multi-button games controllers to play a soccer video game. **Above (right)** Children having fun on a giant Pong game at the National Videogame Museum in Frisco, Texas. **Below** Atari 2600 Breakout Game Cartridge contained just 2 kilobytes of ROM storage and are now popular with retro gamers.

The three founders of Apple, Steve Jobs, Steve Wozniak and Ronald Wayne all previously worked for Atari. Jobs and Wozniak helped build Atari's 1976 arcade game, Breakout.

163

Great Inventions We Take For Granted

The progression in computing power and gaming capabilities can be clearly seen from the rise of sound and speech. In 1980, Taito's Stratovox was the first arcade game to feature synthesized human speech. The game reproduced four phrases – "Help me, help me", "Very good!", "We'll be back" and "Lucky". The 2008 PC game, Company of Heroes: Opposing Fronts featured 50,000 lines of speech, more than a typical movie.

The arrival of microprocessors and further computing advances in the 1970s, saw computer arcade games increase in sophistication. Japanese companies, Taito and Namco, revolutionised arcade gaming with fast action and colour graphics. Taito's Space Invaders in 1978, for instance, was the first shooter where the targets - the alien invaders - shot back and also the first game to save and display a high score, prompting players to insert another quarter to try and top the tally. In the US alone, Space Invader games were played in arcades more than four billion times between 1978 and mid-1981. Namco produced Rally X, the first arcade game to feature a musical soundtrack, along with Galaxians and Pac-Man the following year, the latter developed by a 24 year-old Toru Iwatani and its iconic central character inspired by a pizza missing a slice.

More Pac-Man consoles (over 400,000) were built than any other arcade computer game.

Arcade gaming would be hit by the arrival of home computers and games consoles in the late 1970s onwards but continued to innovate. Racing and sports games with realistic controls were developed such as Namco's Rapid River which used a pneumatic power system to buck and shake the two-seater raft in which players raced and steered using a realistic raft paddle. In 1997, Konami invented Beatmania where players acted as DJs mixing songs and beats in time. It sparked a series of rhythm games including Dance Dance Revolution, where players step on pressure-sensitive squares as they light up to dance in time with the music and moves displayed on the screen.

Home Consoles

In the 1960s and most of the 1970s, access to computing was severely limited to those working within the industry. Ralph Baer and colleagues at Sanders Associates developed a simple electronic games console in 1968 nicknamed the "Brown Box" for the brown wood-grain, self-adhesive vinyl covering used to make the prototype look more attractive to potential investors. It could be instructed to play a small selection of games by flipping the switches along the front of the unit. Plugging into a home TV set, the black and white games which included a bat and ball game that inspired Atari's Pong, were controlled by a box with two dials and one button. The console was licensed to Magnavox, who released the system as the Odyssey in 1972 – the first home video games console.

Above left People enjoying some classic games in an arcade. **Above right** 'Game over' screen from a Pac-Man arcade game. **Above (lower right)** Atari 2600 video games and game manuals from circa 1982. **Above (inset)** The Nintendo Famicom video game console designed by Masayuki Uemura was released by Nintendo in Japan in 1983. Short for "Family Computer," the Famicom was Nintendo's first video game console and led to the popular Nintendo Entertainment System (NES).

No.72 computer games

Whilst pioneering devices enjoyed modest success such as Milton Bradley's Microvision – the first handheld cartridge games machine in 1979 and the Atari Lynx – the first colour LCD display handheld console – Nintendo's GameBoy and DS systems dominated the market in the era before smartphones. By 2016, over 948.5 million Nintendo DS machines had been sold worldwide.

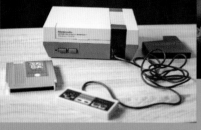

Five years later, Atari diversified into home gaming with the Atari Video Computer System (VCS), later known as the Atari 2600 up against new consoles by Coleco and Fairchild amongst others. With just 128 bytes of internal memory (4.2 billion times less than a modern Xbox One), Atari's console accepted games stored on chunky plastic cartridges containing just 4kb (later 16kb) of memory. Despite the memory limitations, the console and its games, controlled by paddle wheel or joystick and eventually rising to 450 titles, were hugely successful and, along with the arrival of home computers including the Commodore Vic 20 and 64, the ZX Spectrum, TRS-80 and Apple IIe, reinforced gaming as a pastime. For some young programmers, getting their hands on a computer for the first time, they became a source of income as they produced their own homebrewed games which they sold by mail order on audiocassette or floppy disc.

Cartridge-based home consoles became dominated by two main players in the 1980s, Sega and Nintendo, the latter a Japanese playing card manufacturer which had branched out into instant rice meals, taxi cabs and toys and games. In 1980, it produced its first handheld electronic game, the Game and Watch, and followed it three years later with the Nintendo Entertainment System (NES). The company's trump card was games designer, Shigeru Miyamoto, who reeled off a series of classic games in the 1980s that had a profound impact on gaming including Donkey Kong, The Legend of Zelda and Super Mario Bros.

Above (top) First generation Nintendo Game Boy. **Above (center)** Atari 2600 console came with a pair of joysticks and two gaming paddles. **Above (left)** Original Nintendo NES video game console launched in North America in 1985. **Above (right)** A gamer plays Legends of Zelda on the Nintendo Switch console. **Inset (right)** top Child playing Minecraft, a sandbox game developed by Swedish programmer, Markus Persson. **Right** The character Steve from Minecraft.

Great Inventions We Take For Granted

"The Web is more a social creation than a technical one. I designed it for a social effect - to help people work together - and not as a technical toy. The ultimate goal of the Web is to support and improve our web-like existence in the world."

Tim Berners-Lee, Weaving The Web, 2008

the website

The World Wide Web's ubiquity and importance is such that it is increasingly hard to remember and comprehend a world without websites...less than 35 years ago.

On Christmas Day 1990, the very first web page began running on a NeXT workstation acting as the world's first Web server located at the Conseil Européen pour la Recherche Nucleaire (CERN), now known as the European Organization for Nuclear Research in Switzerland. The devisor, Tim Berners-Lee was the son of two early computing luminaries who met whilst working on the pioneering British computer, the Ferranti Mark I. Their son was fascinated by mathematics and science at a young age and whilst studying physics at Queen's College, Oxford University, Berners-Lee built his first computer from a Motorola 6800 processor, TTL gates and the innards salvaged from an old television set which he bought for around $8.

After working as a computer contractor, Berners-Lee was employed at CERN for a period where he developed an information system to keep track of the bewilderingly large number of projects and personnel working at the research centre. Inspired by hypertext links developed and pioneered in the United States by Ted Nelson and Douglas Engelbart, his system produced a vast series of connections between different people and projects which could be navigated by clicking on one hyperlink to progress to another piece of information. He named the software ENQUIRE after an old encyclopaedia he recalled from his childhood, Enquire Within Upon Everything which was devised as a complete guide to all household matters.

Berners-Lee returned to CERN later in the 1980s and proposed a way of producing shared information which could be distributed over computer networks and employed an open architecture, able to run on any computer, regardless of its operating system, and accessible remotely via a computer network. During 1989 and 1990, Berners-Lee developed the various components that would constitute the World Wide Web, collaborating along the way with colleagues such as Robert Cailliau and graduate student Nicola Pellow (the pair would write the first web browser, MacWWW for Apple Macintosh computers in 1992).

No.73 the website

Far left Tim Berners-Lee, British computer programmer and web pioneer. **Top banner** A user types in a website address into their computer's web browser. **This page (left)** Touchscreen icons which launch popular social media sites, apps and ecommerce locations all found as websites on the World Wide Web. **Top right** Researching business data stored on a website. **Center (right)** A website's multiple pages can be accessed easily via clicking on hyperlinks.

Besides the browser, Berners-Lee fashioned three further key elements: HTML (Hypertext Markup Language) to prepare webpages for publication and which gave instructions to web browsers as to how to display them, URLs (universal resource locators which were originally called universal resource indicators) to give each piece of information an address or identity and thus a capability for it to be found and retrieved, and HTTP (Hypertext Transfer Protocol), the protocol or set of rules to enable information transfer across different communications and computer systems.

These were all up and running, albeit in simple form, on Christmas Day, 1990, when Cailliau and Berners-Lee used their computers to perform the first successful communication between a Web browser and server via the Internet. By August 1991, the website was available over computer networks to the public. Contrary to popular belief, the World Wide Web was not an overnight success. It took months of hard lobbying by Berners-Lee, Cailliau and other converts to convince many at CERN as well as other members of the Internet community to come onboard and start using the Web. In June 1993, there was still only 130 websites throughout the world, but momentum built sharply that year, aided by the arrival of the first web browsers capable of handling graphics include Erwise, ViolaWWW and Mosaic.

Berners-Lee and CERN could have chosen to protect, commercialise and exploit the World Wide Web from the start, and potentially made a fortune, but the fervent belief was that for the system to succeed, it needed to be available freely to all. CERN released the project's source code under a general licence making it freely available. Website numbers rose to more than 2,400 websites in 1994 and 51.6 million in 2004 including digital behemoths such as Amazon, eBay and the Google search engine. The Facebook website went live in 2004 followed by YouTube the following year. A further decade on and the number of websites, active and dormant, went past one billion for the first time, according to NetCraft and Internet Live Stats.

Great Inventions We Take For Granted

"I remember Sony Ericsson in 2001 showed off a phone with a clip-on camera. Along with everyone else, I thought 'why would you want a phone with a camera?'"

Jonathan Margolis, technology writer, The Financial Times

the digital camera

Film cameras were all the rage until digital technology offered instant viewing of images with no processing and development delay, no film cost and, with large capacity memory cards, seemingly unlimited photo taking capacity.

The necessity of completing a roll of film before one's images could be processed and printed, prompted American inventor, Edwin H. Land overcome this delay with his Polaroid instant camera which first went on sale in 1948. It contained slim pouches of chemicals and film that developed themselves. Within 60 seconds of taking the photo, you could pull the pouch apart and view your image. Digital cameras pioneered in the 1990s by the likes of Nikon, Sony, Fuji and Ricoh, offered a faster, simpler and ultimately, cheaper way to take and view images instantly. Digicams developed out of the booming electronics industry of the 1960s and 1970s where many engineers and researchers worked on the pressing need for faster, more versatile digital memory for computers and other digital devices. A team working at AT&T Bell Lab, headed by George E. Smith and Willard Boyle, invented the charged couple device (CCD) as a memory component but noted in their 1969 patent how it might find future application as an imaging device.

This was the thought of Steve Sasson, an Eastman Kodak engineer who in 1975 deployed a CCD on a chip at the heart of his innovative digital camera system. Using parts cobbled together from Kodak Super 8 movie cameras along with an audiocassette tape recorded fitted into the machine, Sasson's device weighed nigh on eight pounds and was the size of a toaster. Its lens focused light onto the internal CCD sensor which had a resolution of just 10,000 pixels (0.01 megapixels). The sensor registered the light striking each pixel, measuring the charge which was converted into digital data sent to the tape recorder for storage. Each 100 x 100 pixel black and white photo took 23 seconds to record onto audiocassette and could be viewed by plugging the camera into a monitor or television.

Kodak didn't immediately latch onto Sasson's innovation even after, Bruce Bayer developed the Bayer Colour Filter enabling CCDs to capture colour images. As Sasson later told The New York Times, "They were convinced that no one would ever want to look at their pictures on a television set."

No.74 the digital camera

Although thought of as a 21st Century gadget, the selfie stick was first patented in 1984 by Hiroshi Ueda and Yujiro Mima from Japan. Ueda developed the device after asking someone to take a photo of him in Paris only for the passer-by to steal his camera.

Cameraphones

Mobile phones didn't get digital cameras until the turn of the new millennium with three devices making competing claims for primacy: the Kyocera VP-210 Visual Phone, the Samsung SCH-V200 and the Sharp J-SH04, the latter sold in Japan from November 2000. Its low-resolution 110,000 pixel (0.11 megapixel) image sensor and 256 color display was deemed at the time as both outrageously advanced and potentially unnecessary as the camera could only hold a handful of images and picture messaging was in its infancy. It was only sold in Japan so American phone users had to wait until 2002 for the arrival of the Sanyo SCP-5300, a phone which enabled users to see the results of their photography instantly on screen.

Exponential increases in processing power, advances in memory capacity and the rise of the smartphone has seen camera phone technology progress at an explosive rate. Modern smartphones with 20 megapixel or more image sensors are commonplace (the Xiaomi Mi 9 phone boasts a 48 megapixel sensor), multiple cameras are packed into the same slim device (the Nokia PureView 9, for example, features five) and on board software enables picture correction, customisation and effects only available, less than a decade ago, on a powerful computer workstation. Today, an estimated 3.7 billion digital photos are taken every single day. Apart from giving people the ability to snap any time and all the time leading to the ubiquitous selfie, millions of social media posts and the rise of the concert crowd illuminating the arena with their phones as they try to get a crucial shot, camera phones have also helped break and document many major news stories around the world.

Left page (inset) and above (right) A digital SLR (single lens reflex) camera offers photographers a vast range of shooting and customisation opens viewed on its rear colour screen. **Left page (top)** People snap concert photos using either dedicated digital cameras or the camera in their smartphone. **Top (right)** A telescopic selfie stick enables camera owners to get in the shot. **Right** The latest smartphone cameras boost 20 megapixel or more resolution.

Great Inventions We Take For Granted

Playtime has certainly evolved in the past century with the invention of board games, construction sets and iconic figures which are played with and enjoyed worldwide.

fun and games

Scrabble

The world's most famous word tile game was the brainchild of an out-of-work New York architect, Alfred Mosher Butts, who after analysing various word and number games in the 1930s developed his own board game, allegedly using the preponderance of letters on the front page of The New York Times to determine the letter distribution in the game. Butts named the game, played on a 15 by 15 square grid, Lexiko which he changed to Criss Cross Words, before selling the rights to James Brunot, a Connecticut social worker.

Brunot, his wife, Helen, and their family hand-manufactured sets of the game, renamed Scrabble in 1948, out of a converted schoolhouse in Dodgingtown, Connecticut. The following year the family produced 2,400 sets but lost money. However, interest built and by 1953, the game was selling so well out, especially of Macy's department stores, that the Brunots struggled to keep up with demand and licenced it to a larger games manufacturer. Since that time, Scrabble has gone from strength to strength, with national and international competitions, more than 4,000 Scrabble clubs around the world, and sets available in 29 different languages.

Monopoly

Convention states that George Darrow developed the famous property-accruing board game which he licenced to Parker Brothers in 1935 when demonstrating his home-made set to the company. It featured handwritten Chance cards, houses made from wooden molding offcuts and 10 playing pieces made from charm bracelet tokens: an iron, purse, lantern, race car, thimble, shoe, top hat, battleship, cannon and a rocking horse.

Darrow may have been, at the very least, influenced by an earlier property game devised by actress and stenographer, Elizabeth Magie and patented in 1904, to publicise the negative aspects of capitalism and excessive wealth. Magie's The Landlord Game saw players circle the board, buying properties, railroads and paying rent and avoiding the "Go to Jail" square in one corner. The Landlord Game could be played under one of two sets of rules; the aim under the monopolist rules set was to form monopolies and force other opponents out of the game. Parker Brothers would actually buy the patent from Magie in the same year it acquired rights to Darrow's Monopoly game.

Within a year of Parker Brother's acquisition of Monopoly, they were producing 35,000 copies of the game per week, such was the demand. Darrow had based the games property names on his favourite resort of Atlantic City. Subsequent versions licenced around the world took their own local place names whilst dozens of special editions celebrated specific sports or cities around the world. Some of the UK edition Monopoly sets manufactured by Waddingtons during World War II were fitted with clandestine silk maps of enemy territory hidden in the board as well as real European currency below the paper money and shipped to British prisoners of war in humanitarian aid packages to assist their escape.

No.75 fun and games

There's only one Z in a Scrabble set, but if you use a blank and play the word QUIZZIFY across two triple word squares, you can score a record 419 points, more than many players' entire game scores.

Left page LEGO® minifigure Robin licenced from the famous movie franchise The LEGO® Batman Movie. **Top (right)** A Scrabble set board, tiles and holder. **Top (left)** Wooden Scrabble tiles. A standard Scrabble set contains 100 tiles in varying distributions with E the most common tile, occurring 12 times. **Center (left)** A player picks up the only Z tile in a game of Scrabble. **Bottom (left)** A junior Monopoly game features a simplified rectangular board. Classic Monopoly boards are square. **Bottom (right)** A pewter car playing piece from a classic Monopoly set.

Great Inventions We Take For Granted

There are 915 million different ways you can combine six eight-studded LEGO® bricks.

The moulds used to produce LEGO® bricks are accurate to within two-thousandth of a millimetre (0.000078 inches).

LEGO®

Ole Kirk Christiansen was a carpenter from Billund in Denmark who started building wooden toys first for his sons and then for sale. He set up the Leg Godt – Danish for "play well" company in 1934 which was later shortened to LEGO®. In the late 1940s, the company invested in injection moulding machines, aware that plastics would prove popular in toy-making, and developed its own set of interlocking plastic bricks, named Automatic Binding Bricks with circular nubs on the top which could be pressed into depressions of the bottom of another brick. It is thought that Christiansen and his team were inspired by Kiddicraft Self-Locking Building Bricks invented in 1939 in Britain by Hilary Fisher Page.

Christiansen died in 1952 but his son, Godtfred, continued development of the bricks as a complete "system of play" so that any brick or component could potentially connect to another. A redesign of the bricks internal structure, patented in 1958 made them more robust and initial customer suspicions and disappointing early sales gave way to increasing success, especially when the company partnered with Samsonite to produce and distribute in the United States. In 1964, the company added instruction leaflets to their sets for the first time and made a major switch in materials from cellulose acetate to the more chemically-resistant and physically robust plastic, acrylonitrile butadiene styrene (ABS plastic), the material still used for the bricks today. Four years later, the company opened its first LEGOLAND® amusement park in Denmark, utilising more than 50 million bricks to build exhibits. Today, as the product has diversified into many different lines including the popular Mindstorms robotics kits, the company, still headquartered in Billund, turn out more than 19 billion individual bricks from its factories every year.

This page (above) Classic LEGO® minifigure breaking through a brick wall. **This page (above right)** A range of different LEGO® minifigure heads stacked side-by-side and one on top of each other. **This page (right)** LEGO® minifigures licenced from famous movie franchises including Batman (the Joker, left) and Star Wars (Darth Vader center).

No.75 fun and games

"One evening, I machined a nozzle and hooked it up to the bathroom sink, where I was performing some experiments. It shot a powerful stream of water across the bathroom sink. That's when I got the idea that a powerful water gun would be fun!" Lonnie Johnson, inventor of the Super Soaker.

Super Soaker

Few toys are invented by a fully-fledged NASA scientist, but the Super Soaker water gun is one such example. The son of a World War II veteran, Lonnie Johnson joined the USAF, worked on the Northrop B-2 stealth bomber programme and on NASA's Galileo space probe developing nuclear power and cooling systems whilst working at NASA's Johnson Space Center. An inveterate inventor, Johnson was tinkering with a new refrigeration system valve of his own design when water shot powerfully out of a valve and across the kitchen – giving him an idea for a pressurised toy water pistol with a far greater range than conventional models.

Developing prototypes to entertain his children and USAF colleagues, Johnson eventually partnered with toy company, Larami (later taken over by Hasbro) to take his gun to market. After multiple redesigns, it featured a 40 fluid ounce water bottle which could be easily refilled and a pump action to build pressure of up to 60 pounds per square inch which would fire a constant stream of water up to 42 feet. Initially sold in stores in 1989 and 1990 as the Power Drencher, the gun was renamed the Super Soaker the following year when two million units were sold. Inducted into the National Toy Hall of Fame in 2015, over $1 billion worth of Super Soakers, in many different variants, have since been produced.

Mr Potato Head

George Lerner, a Brooklyn-based toy maker, invented a set of eyes, noses and other facial features which children could press into a potato or another vegetable to make funny faces in the 1940s. Struggling to gain interest from toy manufacturers initially, Lerner managed to convince a cereal firm to package his parts as a giveaway promotion in their boxes before a small toy manufacturer out of Rhode Island, Hassenfeld Brothers, showed interest.

The original Mr Potato Head went on sale in 1952 at a price of 98 cents with four different noses, hands, feet, eyeglasses and a pipe as well as other facial features. Users had to provide their own potato until kits with a plastic potato were introduced in the 1960s. By that time, Mr Potato Head had gained himself a wife, Mrs Potato Head in 1953 followed by a sister, Yam and a brother, Spud. The product sold extraordinarily well, in part due it becoming the first toy advertised on American television, and provided Hassenfeld Brothers with their first nationwide hit on the way to becoming the toy giant, Hasbro. Mr Potato Head's perennial appeal was summed up by it becoming the only licenced toy in the initial Toy Story movies, where he was voiced each time by comedian, Don Rickles.

Above right A classic Super Soaker with refillable water reservoir.
Above Tourists shooting water guns and having fun at Songkran festival, Thailand.
Right Mr Potato Head – a toy icon.

173

Great Inventions We Take For Granted

Turn on and tune in. Radio has provided news, chats, sports commentary and musical backdrops to millions of people's lives for almost a century.

the radio

The invention of radio as a communications medium cannot be simply laid at the feet of one individual. Whilst the name, Guglielmo Marconi is synonymous with the development of radio, he is one of many pioneers that led towards the emergence of radio as a powerful broadcast medium. The discovery and understanding of the electromagnetic spectrum and with it the differing waves that exist along it can be traced to luminaries including Scottish physicist James Clerk Maxwell and German scientist, Heinrich Rudholph Hertz, the latter demonstrating in 1886 via a 'spark-gap' transmitter that rapid variations of electric current could be projected into space in the form of radio waves similar to those of light and heat.

A number of scientists including Nikola Tesla and Alexander Popov investigated radio waves in the late 19th Century. Tesla invented the first radio-controlled device, a model boat, as far back as 1898 at an electrical exhibition in Madison Square Garden. Tesla controlled the four foot long boat's engine and lights using a simple radio transmitter of his own design, wowing the assembled audience. Three years earlier, 21 year old Guglielmo Marconi, whilst conducting experiments, was able to send signals using radio waves over a distance of half a mile on his parent's estate near the Italian city of Bologna. Four years later, he managed to transmit signals from Wimereux near Boulogne across the English Channel, the strip of sea separating England from mainland Europe, followed in 1901, by making the first radio broadcast across the Atlantic Ocean – between Poldhu in the English county of Cornwall to Signal Hill in Newfoundland, Canada.

Marconi continued to develop the use of radio waves as a carrier for telegraph signals which would be adopted by shipping in subsequent years. He was co-awarded the 1909 Nobel Prize for Physics for his work. Others, including Reginald Fessenden focussed on the possibilities of broadcasting sound, both music and speech, wirelessly using radio transmitters and receivers. In 1907, Lee de Forest patented a vacuum tube device called an audion. It amplified the usually very weak

This page An early three band radio is set to medium wave band and tuned into a station via a tuning knob. **Right page** A boy with his mother listens to the radio using headphones in the 1930s. **Right page (inset)** A classic wooden radio set from the 1940s features buttons to alter the wave band and tuning and volume control knobs.

No.76 the radio

In 1923, The Federal Communications Commission ruled that all new radio stations west of the Mississippi River would have the call letter K in their station name, with W reserved for stations to the east of the river.

"The electric waves which were being sent out from Poldhu (Cornwall, England) had travelled the Atlantic, serenely ignoring the curvature of the earth which so many doubters considered a fatal obstacle...I knew that the day on which I should be able to send full messages without wires or cables across the Atlantic was not far distant."

Guglielmo Marconi on his historic transatlantic transmission.

Great Inventions We Take For Granted

> "More than eighty years after the world's first station was founded, radio is still the most pervasive, accessible, affordable, and flexible mass medium available, especially in the developing world."

Bruce Girard, International Association for Media and Communication Research - IAMCR

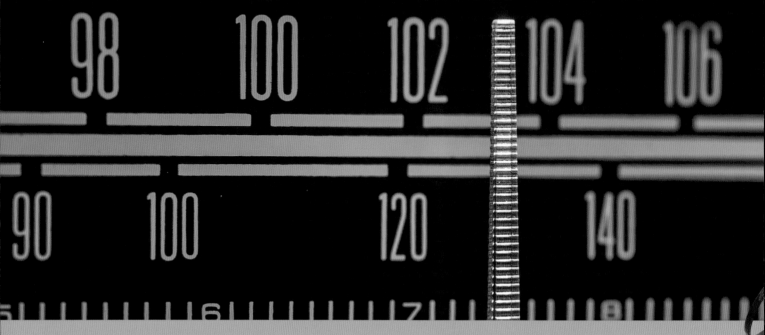

radio signals that arrived at the receiver into a signal clear and loud enough to hear. Vacuum tube-based radio receivers (each colloquially known as 'a wireless') would be widely available in the post-World War I era, evolving from complex, multi-dialled machines for the technically-minded to large pieces of living room furniture with improved reception and sound quality as inventions like the cone loudspeaker were gradually incorporated.

Radio Stations

Whilst experimenters like Charles Herrold in California as early as 1909 were attempting to make audio broadcasts using radio waves to the outside world, it was post-World War I where radio stations and broadcasts really took off. Dutch electrical engineer Hans Idzerda began weekly entertainment broadcasts in 1919 from his PCGG radio station in The Hague, Netherlands. In the UK, the Marconi Company caused a sensation when in June 1920, it broadcast the world-famous soprano singer, Dame Nellie Melba singing operatic arias, Home Sweet Home and the British national anthem.

In the United States. the Westinghouse Company gained a commercial licence to run a radio station in 1920, KDKA, with the intention of producing broadcasts and programmes that would entice consumers into purchasing the company's own radios. On November 2nd, 1920, KDKA made America's first national commercial broadcast. The date was purposefully chosen as the station covered the culmination of the Warren G. Harding–James Cox presidential race, demonstrating graphically how listeners could get the news before they could read about it in newspapers. Commercial radio stations sprang up throughout the United States, mixing with federal and state-run bodies' broadcasts. It all meant that by the end of the 1920s and into the 1930s, a family equipped with a large, cumbersome and usually expensive wireless set, full of bulky vacuum tubes and occupying a corner of the living room much like televisions do now, could listen to a wide variety of broadcasts from weather reports from the US Navy to concerts, religious sermons, lectures and sports broadcasts. In an era before widespread television, the golden age of radio stretched through the 1930s and 1940s providing entertainment and information.

No.76 the radio

Left page A radio tuning display marks a point around the 103.5 MHz point of the FM band. **This page** A boy tunes in to his favourite radio station. **This page (inset)** Competing portable transistor radios from the 1960s. **Above** A Regency TR1 transistor radio.

The Transistor Age
The invention of a replacement to bulky, power-hungry and not overly reliable vacuum tubes came in 1948 when a team at Bell Labs invented the transistor. Used as both a switch and amplifier in circuits, the transistor revolutionised computers and other electrical and electronic products, amongst them the radio. The first transistor radio was the culmination of two companies joining forces, Texas Instruments from Dallas and the Regency Division of Industrial Development Engineering Associates, out of Indiana. Texas Instruments built the transistors which Regency employed in the radio that was launched in 1954 as the Regency TR-1. Compared to earlier radios standing three, four or five feet high, the TR-1 was miniscule at just five inches high and using four germanium transistors inside. It was initially sold at $49.95 (equivalent to some $500 today) yet despite the high price, more than 150,000 were sold. A small Japanese tape recorder manufacturer, Tokyo Tsushin Kogyo, was also working on its own design and introduced its TR-55 in 1955 – the first transistor radio on sale in Asia. Measuring a relatively petite 5.5 by 3.5 by 1.5 inches, but weighing just over a pound, it wasn't quite as light and portable as the company had hoped and subsequent models slimmed down. These included the TR-63 which used an extremely small tuning capacitor and other innovations to shrink its size including an aluminum speaker grille which reinforced the radio's plastic case.

After striking a distribution deal with New York importer, Adolph Gross, the TR-63 became in 1957 the first Japanese radio sold in the US under the Tokyo Tsushin Kogyo's snappier brand name of Sony. It would not fit into regular-sized shirt pockets so the company issued its salesmen with wider-pocketed shirts so they could demonstrate their wares as, "pocket-sized." Sales took off, with more than 100,000 units sold, attracting the attention of other Japanese electrical companies such as Sharp and Toshiba. As they entered the market, transistor radio sales soared. In 1959, some six million Japanese-manufactured radios were sold in the US alone. Radio prices plummeted (as low as $15 as America entered the swinging sixties), increased in quality and could be listened to on the move. Radio stations blossomed with a large upsurge in listenership. People, especially the burgeoning new demographic of teenagers, could finally listen to radio, on their own, away from the bulky family set in the living room. As Apple co-founder Steve Wozniack stated, "My first transistor radio... I loved what it could do, it brought me music, it opened my world up."

Great Inventions We Take For Granted

> "It is by riding a bicycle that you learn the contours of a country best, since you have to sweat up the hills and coast down them."
>
> Ernest Hemingway, US author

the bicycle

Bicycles offer a cheap, convenient and efficient form of transport. A cyclist can travel around three times faster than a walker when both use the same amount of energy. Learning to ride a bike is many people's first taste of the freedom wheeled transport can give.

It is extraordinary to think that the wheel had been in existence for some 5,000 years before someone came up with the concept of placing two, one in front of another, to create a personal conveyance. Whilst claims of sketches and models stretching back to the time of Leonardo da Vinci exist, the first definitive working two-wheeled forerunner of the bicycle was built by a German civil servant, Baron Karl von Drais. His Laufmaschine (known as a Draisienne in France and colloquially as a dandy horse) in 1817 featured two wooden wheels mounted to a frame that also included simple handlebars and a saddle. There were no pedals, chains or gears; the conveyance was propelled by the rider paddling his feet across the ground to push the machine forward.

A short-lived craze sprang up in Europe with many enterprising and intrepid people building their own versions amongst them, the pioneer of photography, Nicéphore Niépce. His model, dubbed a velocipede, featured an adjustable seat whilst other inventions fitted springs from a horse-drawn carriage to improve the ride over the rutted and cobblestone clad streets that existed at the time. Years passed before Scottish blacksmiths such as Kirkpatrick Macmillan experimented with pedals to drive the wheels. Macmillan is thought to have used long rods as linkages to provide rear wheel drive whilst in the 1860s, another blacksmith, Pierre Michaux, this time in France, opted for front wheel drive by joining pedals connected to cranks fitted directly to the axle of the front wheel. A French carriage builder, Pierre Lallement, also fitted pedals to the front wheel, possibly a little earlier in the decade than Michaux, but it was Michaux's Velocipede that helped sparked the second craze for two-wheeled transport that spread through Europe and across the Atlantic into America.

Above People enjoying a bike ride. **Right page (top left)** Cycling champion William Walker Martin sits atop a penny farthing in 1895. **Right page (top right)** A Michaux velocipede with wheel-mounted pedals from the 1860s. **Right page (center left)** A mountain bike with sturdy frame and broad wheels and tires. **Right page (center right)** Two women enjoy a ride on their road cruiser bicycles. **Right page (bottom left)** Men with street bicycles line up at the start of an early cycling race. **Right page (bottom right)** A 20th Century ladies road bike with rear luggage rack and chainguard.

No.77 the bicycle

Great Inventions We Take For Granted

> "To me it doesn't matter whether it's raining or the sun is shining or whatever: as long as I'm riding a bike I know I'm the luckiest guy in the world."
>
> Mark Cavendish, British pro bike racer

The Michaux Velocipede was heavy, weighing over 55 pounds, and with iron 'tires' riding over cobbled streets duly earned its nickname of 'boneshaker'. Without gear systems, one of the only ways one could increase the speed of these bicycles was to fit a larger front wheel. Inventors like Eugène Meyer in France and James Starley in England, created 'penny farthing' bicycles in the 1870s, so named over how the discrepancy in wheel size looked like a large British penny alongside a far smaller British farthing coin. These bikes' front wheels grew up to five feet in diameter with the rider perched high and almost over the front axle. Many found such conveyances hard to handle, the ride position daunting and travel dangerous with the threat of over-the-handlebar crashes an ever-present. Yet, the intrepid Thomas Stevens managed to ride one such bike, a 50 inch wheeled model built by the Pope Manufacturing Company of Chicago, across the across the entire United States - from San Francisco to Boston – between April and August, 1884.

In stark contrast to Starley's penny farthings, his nephew, John Kemp Starley, invented the Rover safety bicycle in 1885 with two very similar-sized wheels mounted to a diamond-shaped frame. The wheels featured tangentially mounted spokes which improved strength. Starley's bicycle also incorporated a recently invented innovation, a set of pedals directly below the rider which rotated a chainwheel and chain linking to a gear set that turned the rear wheel. His machine proved the prototype for millions of bicycles and in the words of mountain bike pioneer, Joe Breeze, "brought bicycling to the masses. The die was cast. A few years later, once the steed was shod with Dunlop's tires, bicycling and its Golden Age craze wrapped around the globe."

A World Of Bikes

Since the invention of the safety bicycle, designers and engineers have spawned a wide range of different two or three-wheeled machines to suit an equally wide range of applications and tasks. Varying body shapes and forms have been trialled, some finding favour including the small framed and chunky-tired BMX bikes for dirt racing and riding, larger wheeled cruiser bikes, popularised by US manufacturer Schwinn, from their introduction in 1933, and the ultra-small-wheeled folding bike style introduced by Dr Alex Moulton in 1962. Moulton's F-shaped open frame bikes were amongst the first to feature wheel suspension, a development later employed on mountain bikes and their hybrid cousins.

Beginning in the late 19th Century with models such as the Fautenil Vélociped, recumbent bikes place their rider in a supine position with their legs and pedals straight out in front of them. Apart from extra comfort and putting less strain on the body, recumbents can be built low so they slice through the air and attain high speeds, as witnessed in 1933, when a pedalled Velocar broke a 20 year old world hour record by averaging 45.05km/h throughout. The following year, the Union Cycliste Internationale – the word governing body of sport cycling, banned recumbent bikes from all competition cycling.

No.77 the bicycle

Mountain Bikes

Bikes had long been used for off-road, trail and downhill riding, terrain that can give a regular road bike a severe pounding. A group of friends in mid-1970s California including Gary Fisher, Joe Breeze, Charlie Kelly and Otis Guy began riding along dirt paths on Marin County's Mount Tamalpais. The group began experimenting and modifying bikes to cope with the challenging terrain. A fruitful route came about through adapting old pre-World War II single speed cruiser road bikes with heavy but robust steel frames and fat balloon tires – nicknaming the bikes 'klunkers'. The Morrow Dirt Club in Cupertino, California were also experimenting in similar fashion and one of their number, Russ Mahon, added a 10 speed gear set as well as powerful and efficient disc brakes to a klunker frame. Whilst many of those involved in these two groups became mountain biking designers or entrepreneurs, it was Joe Breeze who designed the first purpose-built mountain bike frame, debuting it in the Fall of 1977. Breeze produced another nine custom-built frames and bikes by the end of 1978, all featuring strong chrome-moly alloy steel tubing, capable of withstanding the shocks of extreme trail and downhill riding.

Fledgling road bike company, Specialized began developing a mountain bike design with frames produced in Japan then outfitted with components in the United States. The first, the Specialized Stumpjumper in 1981 featured a modified BMX bike stem, handlebars based on motorcycle designs, touring bike cantilever brakes and a 15 speed drivetrain from road cycling. The Stumpjumper weighed 29 pounds, sold for $750 and was advertised as, "It's not just a whole new bicycle, it's a whole new sport," which turned out to be highly prescient with the first UCI Mountain Bike World Cup staged in 1988 and mountain biking's entry into the Olympic games, occurring at Atlanta in 1996.

> "The cyclist is a man half made of flesh and half of steel that only our century of science and iron could have spawned."
>
> Louis Baudry de Saunier, 19th-century French author

Left page (top left) Racing cyclists on the road. **Left page (top right)** Adjusting the derailleur gearset position close to the hub of a rear wheel. **Left page (right)** A row of Citi bikes that form part of New York's bike share scheme. **This page (top)** A mountain biker enjoys tremendous mountain views across British Columbia's rough terrain. **Above** A quartet of children enjoy a ride along a trail.

Great Inventions We Take For Granted

party time

Balloons

It's a grisly beginning for balloons with the first believed to have been fashioned out of pig bladders or animal intestines tied at one end and inflated with air blown in by the mouth. Historians believe the ancient Aztecs made ballons from cat intestines which they proffered to their gods. Natural rubber proved an alternative to animal innards with noted British scientist, Michael Faraday make the first in 1824 for experiments with the newly identified gas, hydrogen. Faraday cut two circles of natural rubber known as caoutchouc and sprinkled flour over the inner surfaces to keep the surfaces separated before sealing the edges together. As Faraday noted, "when expanded by hydrogen they were so light as to form balloons with considerable ascending power."

The first toy balloons were manufactured the following year in London by Thomas Hancock, one of the founders of the modern rubber industry and, several decades later, balloons made of vulcanized rubber, which allowed for a more elastic balloon, were being produced and later prompted The New York Times to state in 1873 that balloons, "will always be an interesting addition to the amusements of popular gatherings." Balloons were imported from Britain and Belgium into the United States in the late 19th Century and featured in Montgomery Ward catalogs at a price of forty cents a dozen. According to balloon historian, Arnold E. Grummer, the first manufacturer of balloons in the United States was the Anderson Rubber Company of Akron, Ohio in 1907. Five years later, came the first long or sausage-dog balloon, credited to another Ohio rubber company, the National Latex Rubber Products of Ashland, Ohio and its proprietor, Harry Ross Gill. Latex, mylar and foil balloons would all follow, enabling printing, customisation and the bewildering amount of choice confronting a consumer who just wants to buy, "some party balloons" today.

Potato Chips

A chef's revenge on a complaining customer in 1853 is reputed to have led to a multi-million dollar industry. The chef in question was George Crum (also known as George Speck) who worked in a lodge near Saratoga Springs frequented by the railroad tycoon, Cornelius Vanderbilt. According to the story, Vanderbilt repeatedly sent his fried potatoes back to the kitchen complaining that they weren't thin enough. Seeking to end the debate once and for all, Crum sent back thin shavings of potato fried to a crisp that had to be handled rather than picked up with a fork.

Whether the story is true or not, the reputation of crispy, thin Saragota chips did begin to spread first to other restaurants

No.78 party time

Left Colorful party balloons filled with helium and tied with ribbons. Potato chips come in a bewildering variety of flavour and ridged or regular textures.

"Unbelievable as it may seem, one-third of all vegetables consumed in the United States come from just three sources: french fries, potato chips, and iceberg lettuce."

Marion Nestle, nutritionist and author of the book, What to Eat, 2013.

Great Inventions We Take For Granted

and later, to food manufacturers who started producing large quantities of chips which were sold mostly loose from bins in grocery stores, leaving a thick layer of broken crisps and crumbs at the base of the bin. A Monterey Park businesswoman, Laura Scudder, had the employees at her potato chip-making factory iron sheets of waxed paper in 1926. These were then turned into sealed bags of chips which stayed fresh and crunchy and were advertised as, "the Noisiest Chips in the World." Other manufacturers began bagging chips using waxed paper and later cellophane and foil helping to create an industry, that according to Research and Markets in 2025 is expected to reach sales in the US of $11.31 billion.

Bendable Drinking Straws

Cleveland-born perennial inventor Joseph Friedmann took the straight straw for sipping sodas and other drinks and gave it a twist. Whilst sitting in his younger brother's store, the Varsity Sweet Shop on San Francisco's 19th Avenue, Friedman watched a young girl struggle to reach the top of the long paper straw nestling in her drink. With an ingenious turn of thought, Friedman fashioned a bendy straw by placing a screw inside the straw before using dental floss to press the straw into the screw thread before removing the screw. The straw now had a corrugated series of ribs along a short part of its length which enabled it to be tilted or bent down at one end, yet still let liquid through. Friedman received U.S. patent #2,094,268 for his "Drinking Tube" in 1937. It took some time to create the machinery necessary to make bendy straws on an industrial scale, but by the late 1940s, Friedmann's Flex-Straw company was selling thousands.

Corkscrew

Whilst the corkless, screw-top closure is becoming more prevalent in wine bottling, hosts and hostesses have long relied on a corkscrew to remove a cork. The first were a simple spiral of metal with a sharpened point fitted to a perpendicular handle made of wood or bone. The earliest known reference to a corkscrew comes from England in 1681 and the writings of plant anatomist Nehemiah Grew who described a corkscrew as a, "steel worme used for the drawing of Corks out of Bottles." In reality, corkscrews date back far further than this point and may have developed from the gun worms used by early musket-wielding soldiers – metal claws with curving tips (the inspiration for the corkscrew's spiral design) used to clear wadding and bullets from the barrels of muskets that failed to fire.

In 1882, German inventor Carl F.A. Wienke, filed a patent for what became an extremely popular and common type of slim, foldable corkscrew he called the "Waiter's Friend," (it later became also known as the Butler's Friend, Wine Key and Sommelier's Knife). It ingeniously used the rim of the wine bottle's neck for leverage to help ease the pull of the cork upwards and out of the bottle. Six years later, H.S. Heeley secured a British patent for Heeley's Double Winged Lever Corkscrew A1, the first in a long line of twin-levered corkscrews which use a simple but effective rack and pinion gearing system to increase leverage. Pushing down on both wings or handles tends to remove even the most stubborn cork. A similar version of Heeley's device was patented in the United States by an Italian, Dominick Rosati in 1930.

No.78 party time

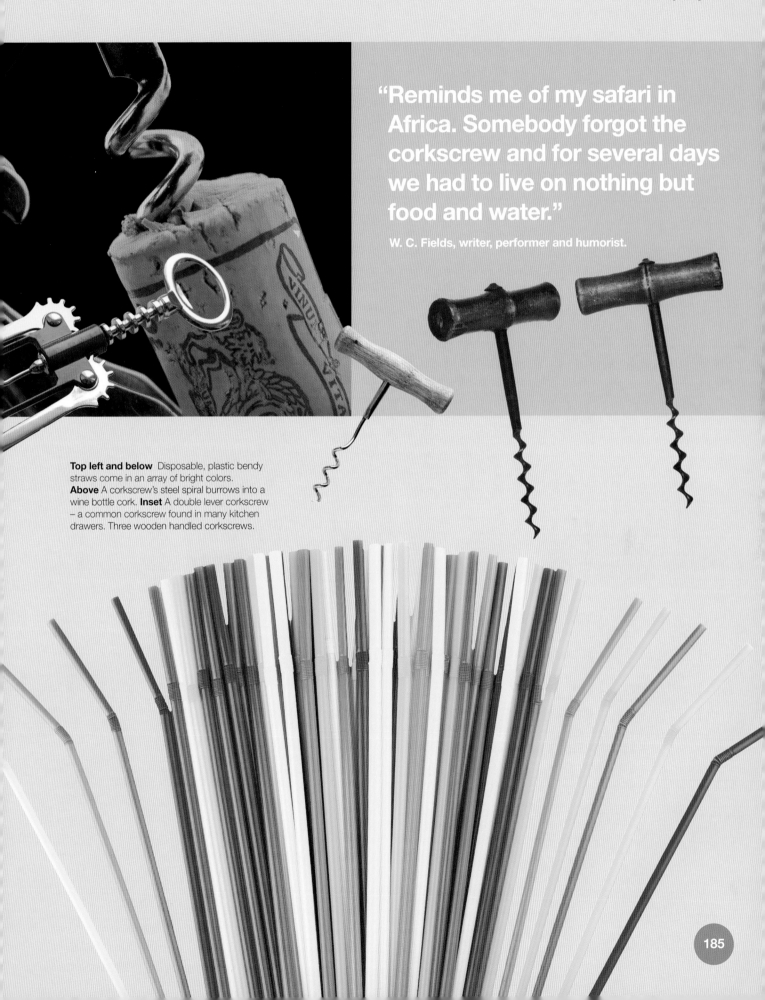

"Reminds me of my safari in Africa. Somebody forgot the corkscrew and for several days we had to live on nothing but food and water."

W. C. Fields, writer, performer and humorist.

Top left and below Disposable, plastic bendy straws come in an array of bright colors.
Above A corkscrew's steel spiral burrows into a wine bottle cork. **Inset** A double lever corkscrew – a common corkscrew found in many kitchen drawers. Three wooden handled corkscrews.

Great Inventions We Take For Granted

Sound travels as a series of vibrations in the air and for many centuries, scientists and scholars have been fascinated by how it worked.

sound recording and playback

Advances in technology and understanding in the 19th century led to the first primitive sound recorders in the 1870s and 1880s. Since that time, dozens of inventions and innovations have enabled increasingly high quality sound to be recorded and replayed with increasing levels of flexibility and convenience. As the innovations progressed, they helped spawn a gigantic and global recorded music industry.

"Mary had a little lamb." With these words, the first line of a nursery rhyme, the great American inventor, Thomas Alva Edison ushered in the age of sound recording. Others has tinkered with capturing sound but Edison's system - the phonograph - first trialled in 1877, was the first which could both record and play back sound. Edison's improved phonograph in the 1880s featured a cylinder coated in wax (the first used a layer of tinfoil) driven by a mechanical wind-up motor. This rotated the wax cylinder as someone spoke into the mouthpiece. The sound vibrations of the voice caused a recording needle to make indentations in the wax. When playback was required, a different needle would travel as the wax cylinder turned, vibrating to produce sounds out of an acoustic horn.

Orders poured in for the cylinder phonograph, but each waxed cylinder in the early years had to be produced individually. In 1887, Emile Berliner, produced recordings on flat disks of zinc which could be processed and then photoengraved for use as a master disc from which other copies of the recording could be pressed and made. Berliner's gramophone records debuted on public sale in 1892 starting out with five inch diameter discs made of hard rubber before he later switched to using shellac – a plastic-like substance produced from resin secreted by the female lac insect. Shellac was also used to produce many of the rival phonogram records that were produced whilst Edison, who switched from cylinders to almost quarter of an inch thick Diamond Discs preferred Amberol (an early plastic) wood flour or china clay coated in resin. Record speeds also varied until 1925 when 78.26 RPM was selected as a standard due to the ease with which certain electrical motors could be stepped down by a toothed gear. These '78s' with up to five minutes playing time per side on a 12 inch disc, became the pre-eminent recording medium until after World War II.

No.79 sound recording and playback

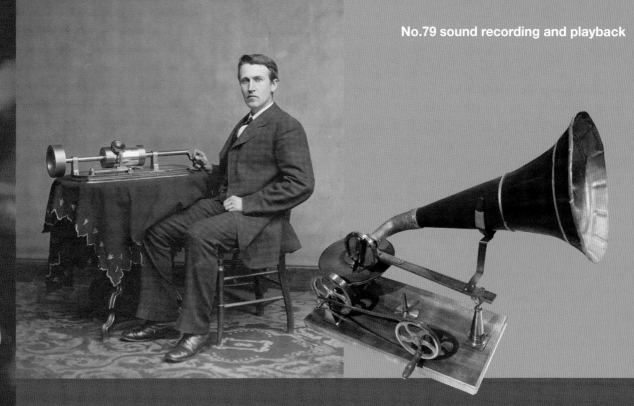

"Mr. Thomas A. Edison recently came into this office, placed a little machine on our desk, turned a crank, and the machine inquired as to our health, asked how we liked the phonograph, informed us that it was very well, and bid us a cordial good night."

Scientific American, December 22, 1877 issue

Left page A mixing desk in a sound recording studio.
This page (far right) Mountain Chief, a leader of the Blackfoot native Americans is recorded by ethnologist, Francis Densmore in 1916.
Right An excerpt of Edison's 1880 patent for a phonograph.
Above Thomas Edison who explored the possibilities of sound transmission with microphones and telephone transmitters as well as the phonograph **(right)**.

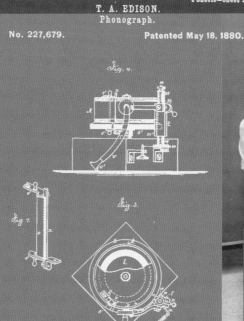

187

Great Inventions We Take For Granted

"Vinyl is the real deal. I've always felt like, until you buy the vinyl record, you don't really own the album. And it's not just me or a little pet thing or some kind of retro romantic thing from the past. It is still alive."

Recording artist, Jack White

LPs and Singles

Hungarian-American engineer Peter Carl Goldmark was head of CBS laboratories which developed the long playing (LP) record. These featured microgrooves holding the sound information etched into vinyl discs at a rate of 250 per inch as compared to a 78's 80 grooves per inch. The discs were spun at 33 and a third RPM allowing far more sound to be recorded and played back on each side of the disc – a little over 20 minutes per side on the initial pressings. Columbia launched their new 10 inch and 12 inch long playing records in 1948, along with a low-priced turntable to play them on. Initial recordings including classical favourites such as Mendelssohn violin concertos and songs from the Broadway musical, South Pacific.

The following year, RCA launched the seven inch single which span at 45 RPM. Both formats began battling with each other for primacy, with some record companies issuing albums as sets of 7 inch records to complete with LPs. However, these two new formats helped hasten the demise of the older 78s whose production dwindled in the late 1950s. By the end of the 1950s, 12 inch LPs were available in stereo and the 1960s would see both seven inch singles and 12 inch albums flourish, with the latter providing a medium for artistes to expand their horizons and release longer pieces of music or collections of tracks with themes. Some record companies pushed the capacity of LPs getting 50-55 minutes duration out of a of a single record; an artist who had more material to release would resort to double or, even triple album releases.

Record sales boomed especially in the US, Britain and Western Europe. Seventy-two million records were sold in the UK, for example, in 1960 but this figure more than doubled by the mid-1970s. In the United States, record sales boomed to $600 million by 1960 but by 1970 had passed the $1.6 billion mark and in 1978, a staggering 762 million records and audio cassettes were sold in the US, worth in excess of four billion dollars.

Vinyl Revival

Whilst vinyl records co-existed relatively peacefully with audio cassettes, the arrival of the compact disc signalled a major downturn in its fortunes. CDs were heralded as offering advanced digital sound, were far more durable and carried far less wear and tear issues such as warping or suffering surface noise, pops, crackle and static issues associated with vinyl. They were smaller, easier to store and users could select tracks in any order more easily than dropping the stylus down onto mid-album. CDs overtook vinyl in sales in 1988 and the following year, Sony Records halted vinyl record production in-house, one of many record companies to completely or largely abandon vinyl which became mostly restricted to clubs, DJs, indie releases and hobbyists. An upturn in vinyl's fortunes began in the early 2010s with a new younger generation of millennials keen to experience the format. Sales in 2018 topped 16.8 million records in the US – a fraction of records in their heyday but a healthy 12% of all album sales, according to Billboard.

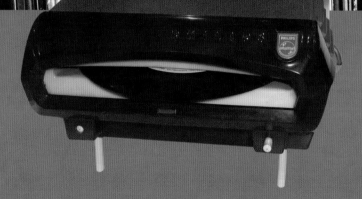

Vinyl Driving / Records On The Road

The LP's inventor also initiated one of the more bizarre trends in in-car audio – a fitted record player under the dashboard. Peter Goldmark, acting on his son's observations that drivers had to listen to whatever available radio stations were broadcasting, developed the Highway Hi-Fi which Chrysler ordered 18,000 units of to be fitted into a number of its car ranges in 1956 including the Thunderbird at a cost of $200 (equivalent to around $1800 today). The unit played newly-designed seven inch records recorded at 16 and two-thirds revolutions per minute so that each record could hold 45 minutes of music or spoken word radio documentaries such as The Battle of Gettysberg. Performance and sales proved poor but it didn't stop other systems such as the Auto Mignon in 1960s Britain (the four members of the Beatles were each said to own one) or Chrysler from making a second attempt with the RCA Victrola miniature jukebox. This 1960 model held up to 14 regular seven inch singles giving a total of up to two and a half hours of playing time but was discontinued the following year.

The average LP has about 1,500 feet of groove on one side. A record stylus's point of contact with the groove of a record covers an area less than two millionths of a square inch.

Left page (top) A 12 inch record is played on a turntable.
Left page (left) An antique jukebox containing multiple 7 inch singles which could be selected by push button control.
Above (right) The Phillips Auto-Mignon – an under-dash mounted record player for cars invented in 1960.
Above A large LP record collection stored on a shelf.

Great Inventions We Take For Granted

US Army Signals Corp major, John 'Jack' Mullin obtained a German magnetophone and 50 reels of tape during the allied occupation of Germany at the end of World War II. He gave two demonstrations of the device in Hollywood in 1947, impressing Bing Crosby who invested $50,000 in Mullin and a small six-man company called Ampex to produce a high quality recorder suitable for radio broadcasts. Crosby became the first major music star to make master recordings on tape whilst Mullin, when he died in 1999, was buried with a reel of magnetic tape.

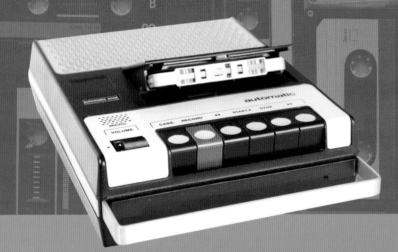

There is 443 feet of magnetic tape in a C90 (90 minute duration) compact cassette.

Magnetized Music

The Danish inventor Valdemar Poulson surprised visitors to the 1900 Paris Exposition by allowing them to record their voices on magnetized pieces of piano wire. Poulsen's Telegraphone had built on the work of New Jersey engineer, Oberlin Smith who had discovered how materials could be magnetized to store sound data. Magnetic wire had limited usage as the wiring had to be permanently fitted within the recording machine but in 1928, German engineer Dr. Fritz Pfleumer made a significant breakthrough when managing to bind ferric oxide powder to thin strips of paper.

Plastic tape bonded with a coating of ferric oxide powder soon followed. The ferromagnetic properties of the tape meant that it could be permanently magnetized when run through the magnetic field of a recording head which receives an electric signal representing sound. This magnetic imprint on a tape stays, unless deliberately erased, and when passed through a playback head, induces a copy of the original electric signal which can be converted back to sound.

In 1935, after years of development, German companies, AEG and BASF (part of IG Farben) collaborated to produce the first successful magnetic tape recorder - the magnetophone - which used plastic magnetic tape. Further refinements led to reel to reel magnetic tape recorders offering the highest audio fidelity in the late 1940s onwards, becoming a standard format for radio broadcasts and in recording studios.

Compact Cassette

A scaled down reel of tape inside a plastic case was developed by the Dutch electronics firm, Phillips in 1963. Whilst most music lovers in the 1960s and 1970s preferred to listen to vinyl at home, compact cassettes gained favour in many motor vehicles, especially outside of North America, where William Powell Lear, inventor of Lear Jet business jets, pioneered the 8-track format. Compact cassette recorder-players allowed millions of ordinary people to record for themselves for the first time, spawning millions of mix tapes and home demos and the first complaints from the recording industry of piracy due to home taping. In 1979, cassettes were on the move not just in cars but also on the belts of pedestrians with the arrival of the small, lightweight, battery-powered Sony Walkman. Designed by Sony's Norio Ohga, the first model, the TPS-L2 weighed 14 ounces and, like most subsequent personal cassettes recorders played its music through on-ear foam-cushioned headphones. Hundreds of millions of machines from Sony and other manufacturers were sold. Sony only stopped production in 2010, long after cassette sales had peaked, and digital music formats had taken over.

No.79 sound recording and playback

"...we are looking at billions of streams on an almost daily basis. If you look at the raw numbers of people who are streaming, I think you could now argue that music has never been more popular."

Paul Smernicki, Universal Records' head of digital 2000-16.

Digital Audio

A wide range of different digital audio file types now exist – from FLAC and WAV to Ogg Vorbis and AAC – but MP3 pioneered and popularised the use of digital files to store sound and music. Developed from the 1980s onwards by the Fraunhofer Institut in Germany, MP3 uses lossy compression where sound waves masked by louder waves and those outside of human hearing range are permanently removed in an attempt to greatly reduce the size of a digital file. MP3s were frequently 8-12 per cent the size of the original sound files making storage and transmission over a computer network a lot faster.

They were first enjoyed via computers via playback software such as Winamp, before the arrival of dedicated portable music players, the first hailing from South Korea in 1998. The Saehan MPMan F10 contained 32MB of Flash storage, enough to hold just a handful of MP3 tracks. The same was the case with the first US MP3 player, Diamond Multimedia's Rio PMP300. Within three years, Apple launched their first generation iPod digital player. Its 5GB storage, huge at the time, could handle 1,000 tracks with ease. By the end of 2014, Apple had sold 390 million iPods as the shift to listening to digital music on smartphones via streaming or digital downloads became inexorable.

Left page (top) German Wehrmacht portable radio transmitter station, featuring a Tonschreiber B reel to reel tape recorder from around 1939. **Left page (left)** Audiocassette recorder-player with push button control and top ejection. **Left page (center)** A chrome C90 audiocassette. **Above (left)** A stack of standard sized compact discs measuring 4.72 inches in diameter and with a thickness of 0.047 inches. **Above** Sound files can be played on digital audio players **(left and center)** or enjoyed on a smartphone **(right)**.

Great Inventions We Take For Granted

credits

Page 1: hurricanehank (Shutterstock.com)
Pages 2-3: Huguette Roe | Dreamstime.com, Flaticon. Pixel-Shot (Shutterstock.com).
Pages 4-5: Galbiati, Yganko (Shutterstock.com).
Contents: lynea, normallens, Plasma_Studio, redstone, Galushko Sergey, arogant (Shutterstock.com). Alexstar | Dreamstime.com
home header: magic pictures (Shutterstock.com)
the band aid: Claudio Divizia, Scott Rothstein, DHurley, Cagkan Sayin (Shutterstock.com). Cupertino10 (Dreamstime.com).

the candle: Holger Schmidt, Cornelius20, Everett Collection Inc., James Kirkikis, Joingate. (Dreamstime.com)

the egg carton: By Anton Starikov (Shutterstock.com), Alexstar, Yakiv Korol, Huguette Roe (Dreamstime.com). Patent - Inventor Joseph L Coyle, COYLE SAFETY CARTON Co. Patent No. US1895974A, United States.

the tetra pak: Tetra Pak [CC BY-SA 2.0 (https://creativecommons.org/licenses/by-sa/2.0)], https://commons.wikimedia.org/wiki/File:Erik_Wallenberg_inventor_of_Tetra_Pak_first_package.jpg. Tetra Pak boy: Tetra Pak [CC BY-SA 2.0 (https://creativecommons.org/licenses/by-sa/2.0)], https://commons.wikimedia.org/wiki/File:Tetra_Pak_boy_with_Tetra_Classic,_Italy.jpg. Tetra Pak [CC BY-SA 2.0 (https://creativecommons.org/licenses/by-sa/2.0)], https://commons.wikimedia.org/wiki/File:Tetra_Pak_Tetra_Brik_with_lady_1960s.jpg. Josep Curto, Adamlee01 (Shutterstock.com).

the microwave oven: homydesign, monshtadoid (Shutterstock.com). Kiattisak Lamchan (Dreamstime.com). Daderot [CC0] https://commons.wikimedia.org/wiki/File:Micro_Cupol_microwave_oven,_designed_in_1969_by_Carl-Arne_Breger,_Husqvarna,_c._1973_-_Tekniska_museet_-_Stockholm,_Sweden_-_DSC01505.JPG.

the mousetrap: Dana Rothstein, Verastuchelova, Artist Krolya, Mohammed Anwarul Kabir Choudhury (Dreamstime.com). Brockhaus Bilderatlas, [Public domain], https://commons.wikimedia.org/wiki/File:Mouse_trap_advertising_19th_century.jpg.

sliced bread: Teodororoianu, Photoeuphoria, Sergii Kostenko (Dreamstime). Eat Less Bread - UBC Library Digitization Centre [No restrictions], https://commons.wikimedia.org/wiki/File:The_Kitchen_is_the_(key)_to_victory._Eat_less_bread_(12659259124).jpg.

the tea bag: Mohamed Osama, Kooslin, Petrle, Liudmyla Ivashchenko (Dreamstime.com). Sheila Fitzgerald, Africa Studio (Shutterstock.com).

the vacuum cleaner: Zoran Skaljac, Marsia16, Paolo_frangiolli, Sandshack33 (Dreamstime.com). Hoover advert: Licensed under the Creative Commons Attribution-ShareAlike 3.0 License. DC07_Vacuum_Cleaner.jpg, F. Duten [CC BY-SA 3.0 (https://creativecommons.org/licenses/by-sa/3.0)], https://upload.wikimedia.org/wikipedia/commons/5/5e/Dyson_DC07_Vacuum_Cleaner.jpg. credit for IRobot: Photo by CEphoto, Uwe Aranas, https://commons.wikimedia.org/wiki/File:IRobot-Roomba-Top-view-01.jpg.

the zipper: Gideon Sundback photo Wikipedia; Old technical drawing U.S. Patent Office Library; Zip central graphic VectorStock; closeup zip photo Fiveprime .

the toothbrush and toothpaste: Africa Studio, 5 second Studio, Andrey_Popov (Shutterstock.com). Dupons Brüssel [Public domain], https://commons.wikimedia.org/wiki/File:Toothbrush1899Paris.jpg.
Staff Sgt. David Gillespie [Public domain], https://commons.wikimedia.org/wiki/File:Toothbrush_teaching_1.jpg. Science Museum, London, CC BY 4.0 (https://creativecommons.org/licenses/by/4.0)], https://commons.wikimedia.org/wiki/File:%27Indexo%27_finger_toothbrush,_New_York,_United_States,_1901-19_Wellcome_L0058113.jpg.

breakfast cereal: Michael C. Gray, AnnyStudio, P Maxwell Photography, by-studio, matkub2499, Jaiz Anuar (Shutterstock.com). Kellogg's company [Public domain], https://commons.wikimedia.org/wiki/File:BlotterKelloggsCornFlakesAdvertizement1910s.jpg.

the iron: Carlos Yudica, Walter Cicchetti, ivan_kislitsin, Maxx-Studio (Shutterstock.com). Sad iron - https://commons.wikimedia.org/wiki/File:Mrs_Potts_sad_irons,_ca_1899.jpg Author unknown, Location: Boston Public Library, Print Department. No known restrictions. https://upload.wikimedia.org/wikipedia/commons/a/a2/Korean_women-ironing_with_sticks-1910s.jpg. [Public domain]. 1882 Seely Electric Flat Iron US Patent US259054 - HENBY W - Google Patents- electric flatiron.

aluminum foil: profartshop, PHILIPIMAGE, Gulyash, stockcreations, Suti Stock Photo, Tanya_mtv, Sergiy Kuzmin, By Daria Medvedeva (Shutterstock.com).

plastic wrap: AlenKadr, eungchopan, daizuoxin, Swapan Photography (Shutterstock.com) Aspenrock, Salmassara (Dreamstime.com).

the diaper: Sabuhi Novruzov, Sirtravelalot, PeterVrabel, Jerome Scholler, ValeStock, Alba_alioth, Teerasak Ladnongkhun (Shutterstock.com).

the coathanger: Yganko, HomeArt, Lawrey, Michael Kraus (Shutterstock.com).

antibiotics: nokwalai, Kateryna Kon, molekuul_be (Shutterstock.com). Alexander Fleming: Official photographer [Public domain] https://commons.wikimedia.org/wiki/File:Synthetic_Production_of_Penicillin_TR1468.jpg.

the aspirin: Shane Maritch, Vikulin, Kalcutta, Everett Historical, Lumella (Shutterstock.com).

braille: XiXinXing, Inked Pixels, Juan Ci, Andy Shell, zlikovec, Stanislav Samoylik (Shutterstock.com).

the can and can opener: Fishman64, Jeff Wilber, CapturePB, Ken Tannenbaum, Ozgur Coskun, gresei (Shutterstock.com). Thomas Altfather Good [CC BY-SA 4.0 (https://creativecommons.org/licenses/by-sa/4.0)], https://commons.wikimedia.org/wiki/File:TAG_Andy_Warhol_Soup_Can_01.jpg.

the cotton swab: By a_v_d, 89studio, Vladimir Gjorgiev, muratart (Shutterstock.com).

the dishwasher: Leszek Glasner, Everett Historical, TheHighestQualityImages (Shutterstock.com). McClure's Magazine [Public domain], https://commons.wikimedia.org/wiki/File:The_Faultless_Quaker_Dishwasher_(1896_advertisement).jpg.

the electric toaster: morkovkapiy, Phil Reid, fractu1(Shutterstock.com) Toastmaster label pic https://commons.wikimedia.org [Public domain]. Mona Lisa Credit © Maurice Bennett.

the safety pin: redstone, Phodo Design, BW Folsom, TaraPatta, Pixel-Shot (Shutterstock.com). wikimedia.org/wiki/File:Patent_6281.jpg, U.S. Patent Office - inventor Walter Hunt [Public domain].

the sewing machine: Wolfgang Lonien, Panjigally, peych_p (Shutterstock.com).

toilet paper: images72, COLOA Studio, Claudio Divizia, KPixMining, KITSANANAN, Studio Dagdagaz, By Kaichankava Larysa (Shutterstock.com).

velcro: Zephyris [CC BY-SA 3.0 (https://creativecommons.org/licenses/by-sa/3.0)] https://commons.wikimedia.org/wiki/File:Bur_Macro_BlackBg.jpg. Daniel Brasil, janmayra, KucherAV, Gunter Nezhoda, Josep Curto (Shutterstock.com). NASA astronaut Greg Chamitoff, NASA, http://spaceflight.nasa.gov/gallery/images/station/crew-17/html/iss017e011577.html.

spectacles: Belinda Pretorius, Everett Collection, Morphart Creation (Shutterstock.com). Yifang Zhao, Onion, (Dreamstime.com).

superglue: Alexander Sobol, molekuul_be, By wk1003mike (Shutterstock.com). Jim Mcdowall, Ronstik, (Dreamstime.com).

central heating: Oleksandr_Delyk, SpeedKingz, Alex Tihonovs, BOKEH STOCK, Nadiinko (Shutterstock.com). Zts (Dreamstime.com).

the safety razor: File:The Gillette Blade February 1918 p05.jpg: Unknownderivative work: Begoon [Public domain], https://commons.wikimedia.org/wiki/File:King_Camp_Gillette_1907.jpg. Star Safety Razor: Scientific American Volume 97 Number 24 (December 1907). (Public domain), Joe Haupt from USA [CC BY-SA 2.0 (https://creativecommons.org/licenses/by-sa/2.0)] https://upload.wikimedia.org/wikipedia/commons/3/30/Vintage_Kampfe_Bros._Star_Single_Edge_Safety_Razor%2C_Made_In_USA_%2823648169392%29.jpg. https://commons.wikimedia.org/wiki/File:Star_safety_razor_advert,_1907.png, [Public domain].
Antonov Roman, NIKITA TV, sirtravelalot, StockPhotosArt, RedDaxLuma (Shutterstock.com).

the cellular mobile phone: Bowrann, Roman Vukolov, peterfactors, Rawpixel.com, LDprod, POM POM. (Shutterstock.com). Rico Shen [CC BY-SA 3.0 (http://creativecommons.org/licenses/by-sa/3.0/)] https://upload.wikimedia.org/wikipedia/commons/1/1f/2007Computex_e21Forum-MartinCooper.jpg. Bcos47 [Public domain], https://upload.wikimedia.org/wikipedia/commons/9/9e/IBM_Simon_Personal_Communicator.png.

soap: chaosheidi, AlisLuch, Kat Byrd I, Ivonne Wierink, AlenKadr, Morphart Creation (Shutterstock.com).

the refridgerator: TaraPatta, By OZaiachin, Winai Tepsuttinun, Everett Collection, By Alba_alioth (Shutterstock.com).

frozen food: Africa Studio, Everett Collection, Adisa, By Pixelbliss (Shutterstock.com).

the dinner fork: V&A Museum.

the safety match: Thongchai S, Sucharas Wongpeth, Nuntiya, Atul Richardson
(Shutterstock.com). Vintage matches - Evan & Dylan Plumb.

work header: Santitep Mongkolsin (Shutterstock.com).

bubblewrap: Mega Pixel, Unchalee Khun, ILYA AKINSHIN, pook_jun, Shanna Hyatt, Christopher Elwell (Shutterstock.com).

banknotes: Roman Romaniuk, David May (Dreamstime.com). nimon, M. Scheja (Shutterstock.com). National Numismatic Collection,National Museum of American History [Public domain].

the computer mouse: Tamisclao / Shutterstock.com, Katerina Fabianova. Tamara Bauer (Dreamstime.com) Marcin Wichary [CC BY 2.0 (https://creativecommons.org/licenses/by/2.0)]
https://commons.wikimedia.org/wiki/File:Xerox_Alto_mouse.jpg. Picture of the first computer mouse: SRI International [CC BY-SA 3.0 (https://creativecommons.org/licenses/by-sa/3.0)] https://commons.wikimedia.org/wiki/File:SRI_Computer_Mouse.jpg. Michael Hicks from Saint Paul, MN, USA [CC BY 2.0 (https://creativecommons.org/licenses/by/2.0)] https://upload.wikimedia.org/wikipedia/commons/5/5a/Douglas_Engelbart%27s_prototype_mouse%2C_angled_-_Computer_History_Museum.jpg.

the paperclip: Flaticon, wonderfulengineering.com; shareably.net; alarmguardsecurity.ca; Bob Vila.

the sandwich: Olgany, Neydtstock, Laurie Cadman (Shutterstock.com). Thomas Gainsborough [Public domain] https://commons.wikimedia.org/wiki/File:John_Montagu,_4th_Earl_of_Sandwich.jpg. Grilled cheese sandwich https://commons.wikimedia.org/File:grilled_cheese.jpg (public domain).

the barcode: Oleksiy Mark, Anatoly Vartanov, Wedding and lifestyle, Eliks (Shutterstock.com).

the wheel: Vector_creator, Dja65, Georgii Shipin, Babich Alexander (Shutterstock.com). Wheel on horse drawn cart in Mongolia. Alexandr frolov [CC BY-SA 4.0 (https://creativecommons.org/licenses/by-sa/4.0)] https://commons.wikimedia.org/wiki/File:Horse_cart_mongolia.jpg.

the thermos: AlexGreenArt, eightstock, Dmytro Balkhovitin, Food Travel Stockforlife, Vera Petrunina, design56 (Shutterstock.com).

the wheelbarrow: Ljupco Smokovski, Makeev Petr, Valmedia, a_v_d (Shutterstock.com). File:Modeles_brouettes.jpg https://commons.wikimedia.org/wiki/File (Public domain). Sale assisted wheelbarrows - Charlesdrakew [Public domain] https://commons.wikimedia.org..

the elevator: Konstantin Tronin, mk1one, SIAATH, Vladimir ZH (Shutterstock.com). Copie de gravure ancienne [Public domain] https://commons.wikimedia.org.

cat's eyes: Dmitry Naumov (Shutterstock.com). EMFA16 (Dreamstime.com). https://commons.wikimedia.org/wiki/File:LIGHTDOME.JPG, ELIOT2000 [Public domain]. Advert for Percy Shaw company - Reflecting Roadstuds Ltd. Catseye00.jpg https://commons.wikimedia.org.

the lawnmower: Robomow 110 City 2012-06-05.jpg, Slaunger [CC BY-SA 3.0 (https://creativecommons.org/licenses/by-sa/3.0)]. Menzl Guenter, bogdanhoda, topseller, Birgit Reitz-Hofmann (Shutterstock.com). Unknown (Chadborn & Coldwell Manufacturing in Newburgh, New York) [Public domain] https://commons.wikimedia.org/wiki/File:ReelMower.png.

the magnetic compass: Danny Smythe, Sergei Drozd, Triff, Valentina Proskurina, OLESIA_V (Shutterstock.com).

the umbrella: ubonwanu, beeboys, ajt, lynea (Shutterstock.com). Louis-Léopold Boilly [Public domain]
https://commons.wikimedia.org/wiki/File:Passer-payez-Boilly-ca1803.jpg.

traffic lights: Victor Grow, GaudiLab, ESB Professional, Bokeh Art Photo, DutchScenery (Shutterstock.com).
https://commons.wikimedia.org/wiki/File:Pedestrian_LED_Traffic_Light_NYC.jpg Freier Denker [CC0].

the seat belt: Graphic Compressor, TuiPhotoEngineer, 3ffi, nlin.nee, VectorPixelStar, MIA Studio, rook76, By Benoist (Shutterstock.com).

the adjustable wrench: RedDaxLuma, Edward Laxton, asadykov (Shutterstock.com). Creative Tools from Halmdstad, Sweden [CC BY 2.0 (https://creativecommons.org/licenses/by/2.0)] https://upload.wikimedia.org/wikipedia/commons/2/2e/Fully_assembled_3D_printable_wrench.jpg.

the power drill: Dilara Mammadova, cynoclub, RedDaxLuma, Peter Kotoff, Kzenon, rawf8, Kzenon (Shutterstock.com).

pens and pencils: Galushko Sergey, Caron Badkin, kylesmith, Sam72, kamon_saejueng, Luis Carlos Torres (Shutterstock.com). Gorodok495 (Dreamstime.com) László József Bíró [Public domain] https://commons.wikimedia.org.

post-it notes: Aleksei Golovanov, Kindlena, Berk Ozel, Aila Images, Pixel Embargo (Shutterstock.com).

the paper stapler: xpixel, Pressmaster (Shutterstock.com). https://commons.wikimedia.org/wiki/File:McGill_Stapler.jpg, Mikebartnz [Public domain]. Swingline stapler Daniel Manrique (roadmr@entropia.com.mx) [CC BY-SA 3.0 (http://creativecommons.org/licenses/by-sa/3.0/)].

the watch: NDT, Cesare Andrea Ferrari, MIGUEL G. SAAVEDRA, Sergey Peterman, demidoff, John Kasawa, Dmitry Kalinovsky, BlackCat Imaging, Monkey Business Images, Vladimir Trynkalo, aslysun (Shutterstock.com). p134 - Biblioteca Europea di Informazione e Cultura [Public domain]. p137 - Deutsche Fotothek [Public domain]. Museumsfoto [CC BY 3.0 de (https://creativecommons.org/licenses/by/3.0/de/deed.en)]
https://commons.wikimedia.org/wiki/File:Rolex_Oyster.jpg.
Rolex Cosmograph Daytona - Rastapopoulos at English Wikipedia [Public domain].

the postage stamp: EtiAmmos, Daniela Pelazza, By pavila, (Shutterstock.com). Joseph Baum and William Dallas printers for local postmaster, E.T.E. Dalton [Public domain] https://commons.wikimedia.org/wiki/File:British_Guiana_1856_1c_magenta_stamp.jpg. Julian Fletcher (Dreamstime.com).

play header: Rashad Ashur (Shutterstock.com).

the bikini: Everett Collection Inc., Fashionstock.com, Dabldy, Vladimir Zhuravlev (Dreamstime.com).

the tv remote: Lunx (null) (Dreamstime.com). Rasulov (Shutterstock.com). Steve Wozniak - Michael Förtsch [CC BY-SA 4.0 (https://creativecommons.org/licenses/by-sa/4.0)]. zenith flashmatic - Edd Thomas from UK [CC BY 2.0 (https://creativecommons.org/licenses/by/2.0)].

the whistle: Sashkin, Shelly Still Photo, Valentin Valkov, Nailotl, Knorre, Shyamalamuralinath, Marius Rudzianskas (Shutterstock.com). Carved whalebone whistle dated 1821. 8 cm long - John Hill [CC BY-SA 4.0 (https://creativecommons.org/licenses/by-sa/4.0)].

the swiss army knife: Xiaowen Sun, Infocus (Dreamstime.com). LMWH / Shutterstock.com.

the hammock: chippix, Dezajny, Realest Nature, Paulette Janus (Shutterstock.com).

the headphones: Jane Kelly, Javier Rosano, dubassy, vlabo, JV Korotkova, Yulai Studio (Shutterstock.com).

the lipstick: Jacob Lund, Juri Pozzi, Re_sky, imagehub, gresei (Shutterstock.com).

the newspaper: RossEdwardCairney, Stocksnapper, industryviews, Everett Historical, Marzolino (Shutterstock.com). The New York Times [Public domain].

the little black dress: Wiktoria Matynia (Shutterstock.com).

the credit card: Teerasak Ladnongkhun, Jacob Lund, A. and I. Kruk, Kunal Mehta, Nattakorn_Maneerat,Oliver Hoffmann (Shutterstock.com).

computer games: Thanaphat Kingkaew, REDPIXEL.PL, CTR Photos, Atmosphere1, Dlogger, Tinxi, robtek, By Pit Stock, omihay, Pabkov, emodpk (Shutterstock.com). Spacewar! - Joi Ito from Inbamura, Japan [CC BY 2.0 (https://creativecommons.org/licenses/by/2.0)]. Large Pong game: Nelo Hotsuma [CC BY 2.0 (https://creativecommons.org/licenses/by/2.0)]

the website: JMiks, notbad, fyv6561, everything possible, PureSolution (Shutterstock.com). Tim Berners-Lee - Paul Clarke [CC BY-SA 4.0 (https://creativecommons.org/licenses/by-sa/4.0)].

the digital camera: Ververidis Vasilis, Billion Photos, astarot, AlexeiLogvinovich, Zodiacphoto, antb (Shutterstock.com).

toys and games: p170-171 zaidi razak, Ekaterina_Minaeva, Chones, NeydtStock, urbanbuzz, Kamira (Shutterstock.com). p172-173 - Lewis Tse Pui Lung, Salvador Maniquiz, zaidi razak, DenisNata, NattapolStudiO, Nicescene (Shutterstock.com).

the radio: T.Dallas, Everett Collection, Plasma_Studio, Volis61, Everett Collection, SeDmi, Laborant, Jim Pruitt (Shutterstock.com). TR1 Transistor Radio: Joe Haupt from USA [CC BY-SA 2.0 (https://creativecommons.org/licenses/by-sa/2.0)].

the bicycle: Chompoo Suriyo, HodagMedia, normallens, nullplus, Everett Collection, Istomina Olena, Gilang Prihardono. (Shutterstock.com). 2nd page- ChiccoDodiFC, VGstockstudio, MNStudio, SimonaKoz, Monkey Business Images (Shutterstock.com).

party time: Oleg Elkov, Danny Smythe, Yuliia Kononenko, Jiri Hera (Shutterstock.com). 2nd page: Melinda Nagy, Francesco83, Dudaeva, Anatoliy Sadovski, DenisMArt, Strekoza64 (Shutterstock.com).

sound recording and playback: 186-187 Maffboy, arogant (Shutterstock.com) Harris & Ewing [Public domain]Edison - Levin C. Handy (per http://hdl.loc.gov/loc.pnp/cwpbh.04326) [Public domain] https://upload.wikimedia.org/wikipedia/commons/0/03/Edison_and_phonograph_edit1.jpg. https://commons.wikimedia.org/wiki/File:Frances_Densmore_recording_Mountain_Chief2.jpg. p188-189 hurricanehank, photopixel, Four Oaks, Riki1979 [Public domain].

Every effort has been made to trace the copyright holders of all images in this book. But some have proved unreachable. In order for errors and omissions to be corrected in any subsequent edition of this publication, please contact Moseley Road Inc.